Library
Brevard Junior College
Cocoa, Florida

SOVIET WRITINGS
ON
EARTH SATELLITES AND SPACE TRAVEL

SOVIET WRITINGS
on Earth Satellites and Space Travel

Essay Index Reprint Series

 BOOKS FOR LIBRARIES PRESS
FREEPORT, NEW YORK

Copyright © 1958 by Am-Rus Literary and Music Agency
Reprinted 1970 by arrangement with Citadel Press, Inc.

STANDARD BOOK NUMBER:
8369-1719-7

LIBRARY OF CONGRESS CATALOG CARD NUMBER:
78-117846

PRINTED IN THE UNITED STATES OF AMERICA

Contents

Part I: From Earth Satellites to Interplanetary Travel
by Ari Sternfeld, Winner of International Incentive Award in Astronautics

Preface 13

From Legends to Science of Space Travel 15

Space Ships 19

Man in Outer Space 40

Artificial Satellite Orbits and Observations 61

Utilization of Artificial Satellites 84

On the Space Ship 101

Space Voyages 106

Artificial Satellites of Solar System Bodies 121

Conclusion 149

Part II: The Sputniks

Space Travel Initiated 155
 by Academician V. Ambartsumyan, Byurukan Astrophysical Observatory

From Man-Made Satellite to Moon-Bound Voyages 159
 by Professor V. Dobronrarov, Doctor of Science in Physics and Mathematics

Earth Satellites and Geophysical Problems 163
 by Alexander Obukhov, Director of the Institute of Atmospheric Physics

6   Contents

Sputnik II   **166**
   Report in Pravda, *November 13, 1957*

Life in Sputnik   **180**
   by P. Isakov, M. Sc. (Biology)

Conducting Optical Observations   **189**
   by Professor B. Kukarkin, vice-president of the Astronomical Council of the USSR Academy of Sciences

What Sputnik I Disclosed About Radiowave Propagation   **192**
   By Professor A. Kasantsev, Director of Technical Sciences

Russian Air Force Views on Sputniks   **196**
   by V. Aleksandrov, G. I. Pokrovsky, V. Grenin, and V. Kaznevsky (engineers and physicists)

Soviet Sputniks and Radio Electronics   **206**
   by Academician A. I. Berg

What Sputniks I and II Disclosed About Outer Space   **211**
   Report in Pravda, *April, 1958*

Sputnik III in Flight   **225**
   Soviet Press Release, *May 29, 1958*

Design of the Third Soviet Sputnik   **235**
   Report in Izvestia, *May 20, 1958*

Solar Battery on Sputnik III   **240**
   by M. S. Sominsky, vice-director, Institute of Semi-Conductors, Academy of Sciences, USSR

Model Comparison   **243**
   by Ari Sternfeld

List of Illustrations

Fig.
1. Diagrammatic view of the solar system **20**
2. Comparative dimensions of the Sun and planets **20**
3. How a satellite could be launched horizontally **21**
4. Comparison of the Earth's gravitational pull with illumination **23**
5. Paths to be followed by space ships **24**
6. A liquid fuel rocket **26**
7. A two-stage rocket **27**
8. First Soviet artificial Earth satellite **29**
9. Probable design of an artificial satellite **33**
10. One of the possible designs of an orbital rocket **37**
11. Space ship designed for round-the-Moon trip **48-49**
12. Variations in the weight of a body during interplanetary flight **50**
13. Creation of artificial gravity on the space ship **52**
14. An artificial satellite can travel only in a plane passing through the Earth's center **62**
15. Diagram of satellite's apparent speed with respect to an observer **70**
16. Enlargement of the globe's visible segment with altitude of satellite's flight **71**
17. Trace left by carrier rocket on photographic plate during prolonged exposure **74**
18. Diagram of first artificial satellite's flight during slightly more than one of its trips around the Earth **76**
19. Diagram of first artificial satellite's travel during one day **77**
20. Dawn, day, evening twilight, and night on Earth satellite **82**
21. View of the Earth's surface from an altitude of 225 kilometers **85**
22. Flight around Mars in two years **110**
23. Flight to Mars on a semi-elliptical trajectory **113**

8   List of Illustrations

24 Flight to Venus along a semi-elliptical trajectory   115
25 A flight to Venus along a semi-elliptical path will take more time than a flight to the more remote Mercury   119
26 Speed-up method of surveying the Moon's surface   128
27 An orbital space ship trajectory which will permit observing a wide zone of the Moon hemisphere hidden from our view   137
28 Possible orbital space ship trajectory for observation of the Moon   138
29 Possible alternate trajectories of an orbital space ship travelling on the Earth-Venus route   140
30 Routes of orbital artificial planet-space ships which return automatically to Earth after varying periods   142
31 Satellite above ionosphere (direct, reflected reception)   193
32 Satellite above ionosphere (slanted reception)   194
33 Satellite below ionosphere (reflected reception)   194
34 Satellite near F_2 Layer Maximum Ionization (round-the-world echo)   195
35 Trajectory of rocket-plane's flight   197
36 Design of a liquid-jet engine   201
37 Diagram of a space ship design   203
38 Schematic diagram of Sputnik III   237
39 Diagram of separation of satellite from carrier rocket   238
40 Schematic diagram of solar battery   241
41 Comparison of Soviet and American satellites from viewpoint of energy they possess after being put in orbit   244

List of Tables

I Places and the Dates on Which the First Soviet Earth Satellite (Sputnik I) Appeared Over Them for the First Time in October, 1957   *72-73*

II Time and Speed Required to Reach Other Planets of the Solar System   *118*

III Characteristics of Two Artificial Satellites of the Moon   *125*

IV Zero Circular and Parabolic Velocities for Planets and the Sun   *129*

V Sidereal (Star) Circuits Periods of Zero Artificial Satellites of Planets and the Sun   *129*

VI Characteristics of Satellite Space Stations of Planets and the Sun   *131*

VII Spheres of Attraction of Planets (According to Tisserand)   *132*

VIII Launching of Artificial Planets in Semi-Elliptical Trajectory   *133*

IX Main Characteristics of Orbits of Orbital Artificial Planet-Space Ships Which Return from Outer Space to Earth after Intervals of Not Over Five Years   *144-145*

Table of Equivalents

To convert	to	multiply by
millimeter	inch	0.0394
centimeter	inch	0.3937
meter	foot	3.2808
kilometer	mile	0.6214
kilometers per hour	miles per hour	0.6214
meters per second	feet per second	3.2808
meters per second	miles per second	0.00062
square centimeter	square inch	0.155
square meter	square foot	10.7639
cubic centimeter	cubic inch	0.061
kilogram	pound	2.2046
metric ton	U.S. short ton	1.1023
	U.S. long ton	0.9842
degrees Centigrade	degrees Fahrenheit	9/5 and add 32

One astronomical unit (A. U.) is 150,000,000 kilometers or 93,003,000 miles, the average distance from the Earth to the Sun.

Atmospheric pressure measured by mercury barometer is rated according to the height of the instrument's mercury column in millimeters or inches. Ordinary height at sea level is 30 inches (760 millimeters), the equivalent of one atmosphere. Air pressure at sea level is normally 14.7 pounds to the square inch.

part one

From Earth Satellites
to
Interplanetary Voyages

BY ARI STERNFELD
Winner of International Incentive Award in Astronautics

Preface

When the first artificial Earth satellite (Sputnik I) was launched in the USSR on October 4, 1957, mankind entered the age of interplanetary travel which was brilliantly forecast by Konstantin Eduardovich Tsiolkovsky at the dawn of our century. A month after this event on November 3, the second Soviet satellite with a test animal (dog) aboard was launched. The prospect now opening before the world is a trial flight to the Moon and the neighboring planets—Mars, Venus, Mercury—at first by unmanned space-exploring vehicles under remote control from the Earth, and then by manned rockets with spacemen crews. Human beings are destined to fly not only to other planets, to their moons and comets, but also to areas near the Sun and, in the remoter future, to the distant stars.

The launching of the Earth's first artificial satellites is a great victory for Soviet science and technology in the peaceful competition between the two systems of capitalism and socialism. This victory was won thanks to the diligent organized labor of Soviet scientists, engineers and workers and to the unprecedented growth of science and technology in our country during the years of Soviet power.

The launching of the second Soviet artificial Earth satellite was dedicated to the Fortieth Anniversary of the Great October Socialist Revolution by its creators—the personnel of scientific research institutes and of designing offices, the rocket testers and factory producers.

In the Soviet Union amateur astronautical groups came into existence over 30 years ago, in 1924. These circles and societies had the aim of investigating problems of jet propulsion and interplanetary flight. They united in common effort all persons interested in the space flight field.

Investigations in the field of astronautics took a new step forward in 1954 when a Joint Commission on Space Travel was organized at the Astronomical Council of the Academy of Sciences USSR and a Tsiolkovsky medal was instituted to encourage astronautical research. The commission coordinates the activity of scientific research institutes engaged in studying the problems that must be solved to ensure the further progress of astronautics in our country.

14 Preface

In 1954 an Astronautical Society was instituted at the Chkalov Central Air Club in Moscow. Certain problems of space flights are also being studied in astronautical circles organized at the higher educational institutions of Moscow, Kiev, Kharkov and other cities.

Interplanetary travel has until recently been considered a purely theoretical problem. But yesterday's dream having now become reality in today's artificial celestial bodies orbiting around the earth, many old problems are seen in a new light.

Exploration of the earth and outer space by means of artificial satellites is a component part of the program of the International Geophysical Year (July 1957—December 1958), a scientific undertaking of unprecedented scope. Nations of the whole world, whose representatives now already confer at annual international astronautical congresses, have a part in tracking the earth's first artificial moons. The International Astronautical Federation brings together the national rocket and spaceflight societies of more than twenty countries; its membership is constantly growing.

It is a matter for the nations themselves to decide to what extent they can direct their energies in creative effort and not destructive war, so that the next steps into outer space can be taken in seven-league boots.

ARI STERNFELD

Moscow, November 1957.

From Legends to Science of Space Travel

FOR MANY a century, space travel was considered an idle dream.

There are numerous legends in which man flies to other worlds or visitors come from those worlds to the Earth. Ancient Greek mythology is especially rich in such legends, as, for instance, the story of Icarus, who fixed feather wings to his back with wax and flew so near the Sun that the wax melted and he fell into the sea and was drowned. Another is the story of Alexander the Great, who wanted to visit heaven in a chariot driven by eagles. A Chinese legend asserts that the Chinese came from the Moon.

In the grim period of the Middle Ages people left the idea of space flight alone, fearing persecution from the Church. The exception was the Indian epic Ramayana, whose main character travelled to the heavens.

During the Renaissance there was a revival of interest in fantasies about flights beyond the Earth, legends being replaced by scientific conjectures as man's knowledge of nature grew.

The first schemes of a technical nature for establishing contact between the Earth and other celestial bodies arose in the 17th century. They were not, however, based on scientific theory.

The English scientist John Wilkins referred to the possibility of space flight in his *Discourse Concerning a New World and Another Planet*. Cyrano de Bergerac, the French novelist, went further. Long before man had learned to fly he spoke of the possibility of using rockets for space travel, going as far as to describe the simplest design for a rocket space ship.

The 19th century saw the appearance of a number of fantastic novels on space travel, some of them absolutely unscientific. For example, Jules Verne's heroes were fired to the Moon from a gun. But the author completely overlooked the fact that his heroes would be crushed to death the moment the gun was fired.

At the beginning of this century H. G. Wells in England and A. Bogdanov, and later A. Tolstoi and A. Belyaev, in Russia wrote fantastic novels dealing with the life of other worlds, which were very popular with the reading public.

A number of novels and stories on space flight were also written by scientists, among them K. E. Tsiolkovsky.

◊ ◊ ◊

Nowadays astronautics, the science of space travel, has an indisputable right to be treated as an equal partner with other branches of science. (The term astronautics has the same implications as does cosmonautics, deriving from the Greek "astron" or star, and "cosmos" or universe, plus "nautics" which means everything pertaining to ships and navigation.)

The history of astronautics is closely related to other fields of scientific endeavour. Astronautics could not exist, for instance, without a knowledge of astronomy or Nicolaus Copernicus' teaching on the structure of the solar system.

Copernicus proved that the Earth is not the center of the universe and that all the planets, including the Earth, revolve around the Sun. Johannes Kepler discovered the laws governing the motion of the planets and Isaac Newton clearly defined the principal laws of celestial mechanics. He also considered the possibility of a missile becoming a miniature "Moon," an artificial satellite of the Earth, and of a body receding from the Earth into infinity.

Copernicus' teaching and the laws of Kepler and Newton are of paramount importance for astronautics, because a spaceship can be looked upon as a kind of celestial body, which will follow a strictly defined path in space and be subject to the same laws as a celestial body.

Astronautics made its appearance as a result of the development of astronomy and rocketry.

If we take a brief look at the history of the rocket we find that in ancient times the Chinese fired off gunpowder rockets for entertainment on big festive occasions, and in the Middle Ages rockets were also used for military purposes.

By the end of the 16th century there appeared drawings and descriptions of multi-stage rockets, and in the mid seventeenth century drawings of rockets equipped with air fins.

Russia learned about rocketry at the beginning of the 17th century as a result of the work conducted by Onisim Mikhailov. The first "Rocket Research Establishment" in Russia was founded in 1680. In the middle of the 19th century it was headed by K. I. Konstantinov, the most prominent rocketry expert in pre-revolutionary Russia, who considerably improved the Russian war rocket. In 1881, N. I. Kibalchich designed a rocket flying machine.

The theory of the motion of a rocket in space was elaborated by the famous Russian scientist K. E. Tsiolkovsky (1857-1935), who also designed the first liquid-fuel rocket.

Among his followers mention should be made of F. A. Tsander (1887-1933) and Y. V. Kondratyuk, who died in 1942.

Such foreign pioneers in astronautics as Robert Esnault-Pelterie (France), Hermann Oberth and E. Saenger (Germany), Robert H. Goddard (U.S.A.) have made a great contribution to the science, as have also such popularizers of space travel as A. Ananoff (France), W. Ley, A. Haley (U.S.A.), Y. Stemmer (Switzerland), E. Burgess and A. Clarke (Britain), H. Gartman (F.R.G.) and the American, British, German and other interplanetary space flight societies.

Goddard's liquid fuel rocket was launched in 1926. The first Soviet liquid fuel rocket of M. K. Tikhonravov's design was launched in 1933.

Rocketry has made tremendous progress since then, as can be seen from the following figures. In the Nineteen Thirties a single-stage liquid-fuel rocket reached the record-breaking height of thirteen kilometers; in 1952 the record was 217 kilometers and in 1955, 288 kilometers.

Naturally, the performance of multi-stage rockets was much better: about 400 kilometers in 1949, a little under 500 kilometers in 1953; while now they rise more than a thousand kilometers. Certainly these figures are not very impressive in comparison with the distances between the earth and other celestial bodies. The distance to the Moon, for example, is hundreds of times as great and the distance to the nearest planet is tens of thousands of times as great. However, the achievements of rocketry are already substantial.

In October and November, 1957, the first Soviet artificial Earth satellites were placed in orbit. Orbital rocket speed is now a reality, that is flight velocity sufficient for transition to a circular or elliptical orbit, or for conversion of a rocket into an artificial satellite of earth. One step more, an increase of 150 per cent in orbital rocket speed, and the space vehicle will escape from the earth's gravitational field and fly out toward the nearest celestial bodies—the Moon, Venus, Mars.

It is a widespread belief that there must be a revolution in technology before man can fly into space. This opinion is erroneous. Space flight is gradually becoming a practical proposition. The successful development of rocketry, remote control, physics and biology gives us every reason to believe that mankind is on the threshold of space travel. Today scientists of many countries are working in this field. Astronautics is the concern not only of specialists but the public at large, and, since the war, astronautical societies have been formed in more than twenty countries.

CHAPTER ONE

Space Ships

The Problem of Escape from the Earth

Take a look at the solar system (Fig. 1), whose boundless expanses will be traversed by the space ships of the future.

The Earth, one of the nine large planets of the solar system, travels at a great speed in airless space along an almost circular orbit around the Sun at a distance of about 150,000,000 kilometers from it. This distance is taken as one astronomical unit. The other eight large planets and a great number of small planets—asteroids—travel approximately in the plane of the same orbit. Comparative dimensions of the Sun and the planets are shown in Fig. 2.

Interplanetary space is limited by the orbit of Pluto, the outermost planet of the solar system, which is separated from the Sun by a distance of nearly 6,000,000,000 kilometers. It is this boundless expanse that space ships will cross, making use of the Sun's attraction or struggling against it, and evading collision with wandering meteors and swarms of asteroids.

What prevents us from launching a rocket into space?

The major obstacle is the force of gravity. Anything on the surface of the Earth is attracted to its center; and not only the Earth but all bodies, from a tiny grain of sand to a gigantic star, have this power

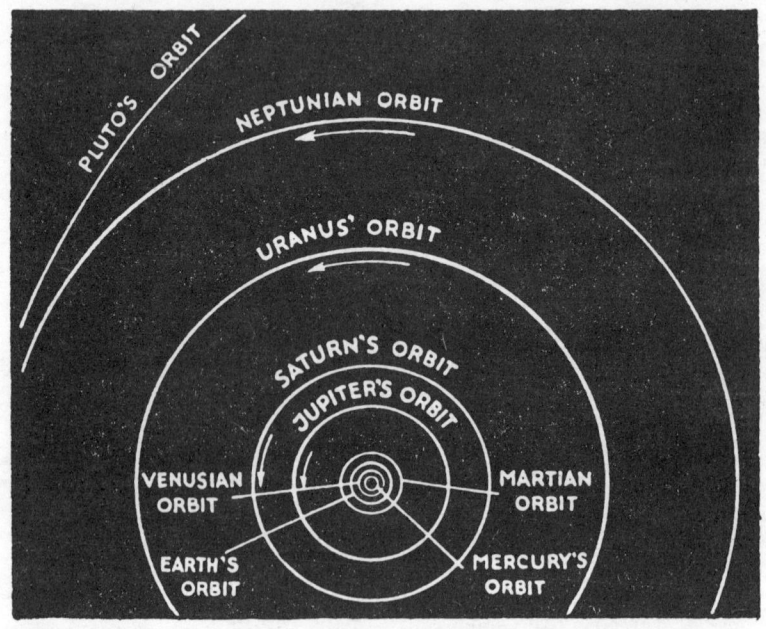

Fig. 1 Diagrammatic view of the solar system

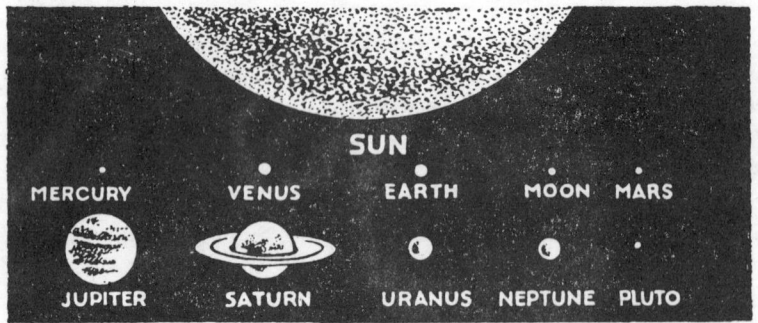

Fig. 2 Comparative dimensions of the Sun and planets

of attraction. All things around us pull at each other, but we do not feel it because their power of attraction is too weak. On the other hand we always feel terrestrial gravity.

Were it not for the force of gravity, there would be nothing left on the surface of the Earth; everything would fly off into space, the Earth would move away from the Sun and the Moon from the Earth. The fact that this force does exist complicates the problem of interplanetary flight.

Is it possible for a rocket to leave the Earth never to return?

Yes, it is. Let us imagine that a horizontal launching site (Fig. 3) has been constructed on a high mountain where air is no longer an

Fig. 3 As the rocket gains speed its range of flight increases and the curvature of its path decreases. Upon reaching circular velocity (the upper orbit), the rocket becomes the Earth's satellite and flies parallel to its surface.

obstacle to the rocket's flight. A rocket launched from such a site at a certain speed would follow a steep trajectory and fall at a certain distance from the mountain. Were the fuel supply and speed of the rocket to be doubled, it would fly farther and its trajectory would be less steep. The speed of the rocket can be increased to a point where the curve of the rocket's trajectory is the same as that of the Earth's surface. As soon as this point is reached the rocket can fly around the Earth and circle it again and again. In this way the rocket can become a satellite of the Earth and, like the Moon, will never fall to its surface.

The lowest speed at which a body can travel round the Earth without falling is called the first astronautical speed or circular speed.

Why does a body not fall to the Earth at this speed?

When an aircraft flies around the Earth along the equator or a meridian it is affected by a centrifugal force which increases in proportion to its speed. This force is in opposition to the pull of gravity and strives to lift the aircraft away from the Earth. This centrifugal force increases with increased travelling speed and is proportional to the square of the speed. At low speeds it is almost imperceptible. For the pedestrian walking along a straight road, centrifugal force amounts to one milligram. For a running man it is intensified many-fold, and for an aircraft flying at record speed of about 2800 kilometers an hour, it reaches one per cent of the plane weight. When circular flight speed reaches 7.9 kilometers per second the centrifugal force equals the force of gravity and neutralizes it, seemingly eliminating its effect on the flying body (this should of course be understood to imply not that the force of gravity has disappeared, but that it has been completely compensated for by the centrifugal force acting in the opposite direction). Were it not for air resistance, an aircraft flying at such a speed would circle around the Earth indefinitely under its own momentum. It would thus become an artificial satellite of the Earth. The body orbits around our planet; in airless space its velocity has become invariable.

And at what speed must a body travel if it is to overcome the Earth's gravity and fly off into space?

To answer this question we must learn something about gravity.

The Earth's gravitational pull, like that of the other celestial bodies, decreases as one moves away from its center. It decreases at the same rate as the brightness of an object diminishes the farther it is removed from its source of illumination, i.e., in inverse proportion to the square root of the distance (Fig. 4). In other words, the Earth's gravitational pull diminishes by a factor of four at a distance twice as great, or by a factor of nine at a distance three times as great, etc.

In order to free a body from a planet's gravitational field the same amount of work has to be performed as for lifting it to a height equal to the planet's radius, assuming that the gravitational force does not change as the body recedes from the planet's center. We can accomplish that by giving the body a certain speed near the surface of the Earth. A body going at that speed would travel along a parabola (Fig. 5), and this is the origin of the term parabolic speed, which is also known as the second astronautical speed or "the speed of escape." At the Earth's surface it is 11.2 kilometers per second.

If the speed imparted to a body is more than the circular speed

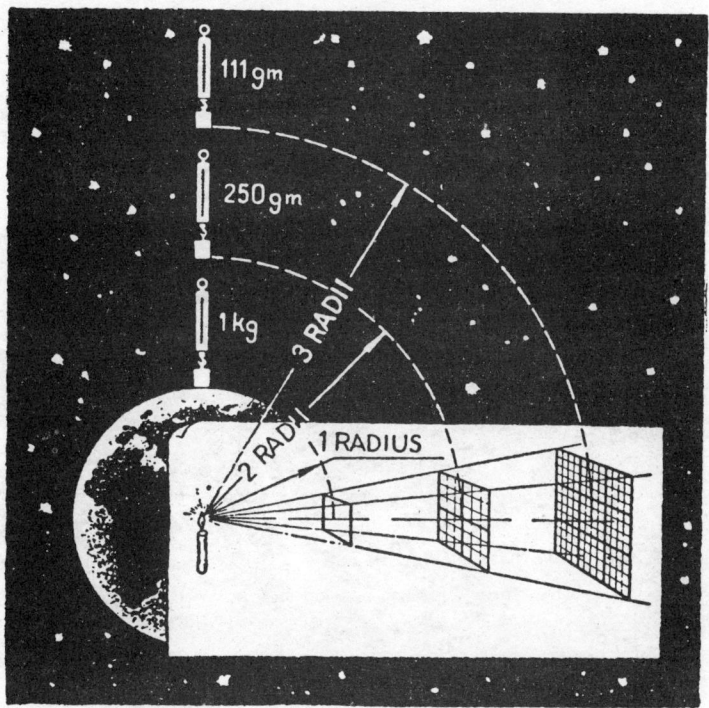

Fig. 4 The Earth's gravitational pull decreases at the same rate as the brightness of objects diminishes the farther they are removed from their source of illumination.

but less than the parabolic speed, it will send the body travelling along an elliptical orbit. Should the speed exceed the parabolic speed, the body will travel along a hyperbola (Fig. 5).

To facilitate calculation I have assumed that the body is attracted only by the Earth. Actually, however, it is also affected by the Sun's gravitational field. Calculations show that a speed of not less than 16.7 kilometers per second is required to set a body free

24 *From Earth Satellites to Interplanetary Voyages*

from the gravitational fields of the Earth and the Sun. This is called the third astronautical speed.

The first of these barriers—achievement of the first astronautical speed—was surmounted with the launching of the first artificial earth satellite. Astronautical science must now make it possible to attain the second, and after that also the third astronautical speeds.

The Rocket, Prototype of the Space Ship

It is generally accepted that any future space ship will be rocket-driven. It will be propelled through empty space by the thrust of the gases ejected from the rocket. Travelling by rocket is quite safe

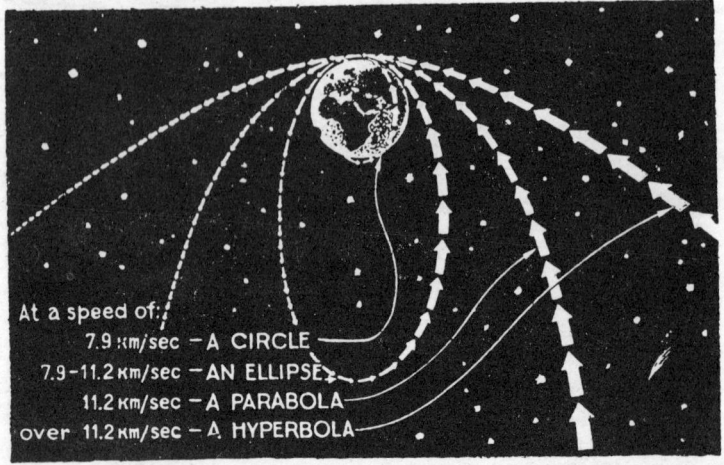

Fig. 5 Paths to be followed by space ships

because the rocket, as distinct from an artillery shell, gains momentum gradually. That is why at take-off the strain on the human body will be comparatively low, causing the astronauts no harm.

Since the speed of a rocket-propelled space ship through the atmosphere will be relatively low, the ship will not meet with much air resistance, and heat due to friction will be insignificant.

The rocket motor will enable astronauts to increase or reduce the speed of the space ship in empty space or change the direction of flight when necessary.

What is the principle of the rocket's motion?

When a rifleman shoots, the rifle recoils. This is the result of the pressure of the gases formed from the burnt gunpowder. The gases press against the bullet and the rifle with equal force but the latter recoils slowly because its mass is much greater than that of the bullet. That is one of the main laws of mechanics, which runs: "To every action there is always an equal and contrary reaction." The motion caused by reaction is known as reaction motion.

The type of rocket with a gunpowder charge that is launched at a carnival cannot be used as motor for the space ship, because of the very high pressure set up by the combustion of gases. To withstand such pressure the rocket must be very strongly built, and consequently very heavy. Moreover, the consumption of gunpowder during flight cannot be regulated, just as the flame of a candle cannot be regulated. One cannot stop the gunpowder burning in order to shut down the motor when necessary.

The liquid-fuel rockets extensively used nowadays are superior to gunpowder rockets in this respect.

Fig. 6 shows a liquid-fuel rocket with two tanks, one containing a propellant (for example, ethyl alcohol), the other an oxidizer (for example, liquid oxygen).

Two pumps driven by a turbine feed the two liquids into a special chamber, where a chemical reaction (i.e., the combustion of liquid fuel) takes place. The gases formed in the process escape from the combustion chamber, producing a recoil that sends the rocket forward.

Both gunpowder rockets and liquid-fuel rockets depend for the steadiness of their flight on air fins and rudders.

But these are of no use once the rocket has left the Earth's atmosphere and emerged into space. What are the astronauts to do if the rocket is deflected from its route? This problem was solved by K. E. Tsiolkovsky, who suggested that rudders be placed in the stream of gas emitted from the nozzle in order to change the direction of the rocket's flight in empty space. This can be done by mounting the engine on swivels instead of fixing it rigidly in the rocket framework, in order that the engine's alignment relative to the rocket's axis of symmetry can at a certain time be shifted.

On what factors does the rocket's speed of travel depend?

In empty space, beyond the reach of gravitational fields, the velocity that can be attained by a rocket depends on the speed with which the gases leave the nozzle and the amount of fuel consumed. In view of this, use will be made of those fuels which produce the greatest possible exhaust velocity, for example oxygen and hydrogen. However, hydrogen is very light, even if condensed into liquid form, and requires more spacious tanks than other propellants. Moreover its boiling point

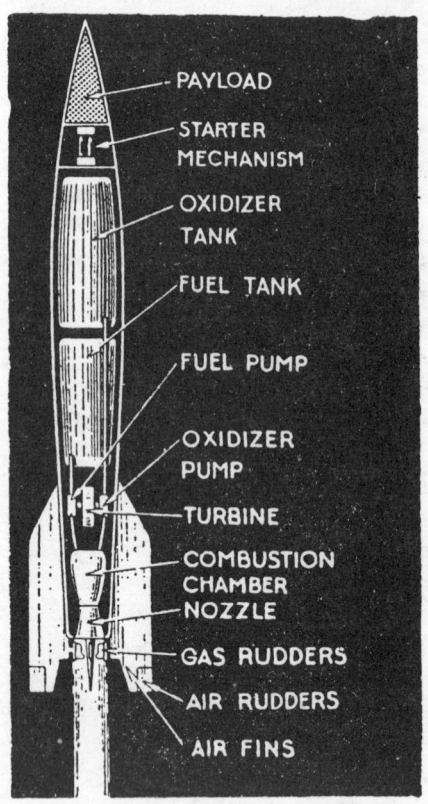

Fig. 6 A liquid fuel rocket

is $-253°C$. Nitric acid and hydrazine (a chemical combination of nitrogen and hydrogen; colorless, poisonous liquid with unpleasant odor) are more economical, inasmuch as these liquids (heavier than water) are easy to handle and can be contained in small tanks. Other liquid-fuel rocket propellants are kerosene, benzine, turpentine, paraffin, etc., with perchloric acid, hydrogen peroxide, etc., serving as oxidizers.

Thermochemical (conventional) propellants produce exhaust velocities of about 2.5 kilometers per second, but there are good reasons for believing that a speed of four kilometers per second can be reached, which would simplify the problem of constructing a spaceship.

Another method of increasing the rocket's speed and range is to launch it with an auxiliary rocket. When the latter uses up all its fuel it is automatically jettisoned, the touch-down being by parachute. The main rocket is fired only when the auxiliary has done its job,

Fig. 7 A two-stage rocket

that is when it has reached a certain height and speed, and it can therefore climb higher than an ordinary rocket. A rocket of this type is known as a stage- or step-rocket (Fig. 7). By adding more steps (boosters), the speed of the rocket and its range can be increased.

Experiments conducted during the last few years have shown that booster gunpowder rockets are very economical because the thrust they develop is enormous in comparison with their weight. They are likely to be used for the intial projection of the space ship.

To increase exhaust velocity still more, conventional propellants should be replaced by nuclear propellants.

What is a nuclear propellant, and why is it superior to conventional propellants?

Modern physics has succeeded in converting a number of chemical elements into other elements. In certain cases the process is accompanied by a release of atomic energy. A material that produces such energy is known as a nuclear propellant, a small quantity of which contains enormous energy.

The release of atomic energy is a very rapid process, but it is not altogether uncontrollable.

Atomic energy can be used to convert certain liquids (for example, liquid hydrogen or helium) into gas and then to expel it from the rocket.

Nuclear propellant with a liquid or gas is called "atomic fuel."

One must bear in mind that the terms nuclear propellant and atomic fuel are used here only conventionally, since the release of atomic energy and its transfer to an inert body bear no resemblance to the process of burning as we know it.

In an atomic rocket the gases will leave the nozzle at a speed of several dozen kilometers per second, and the greater the exhaust velocity the less fuel is required for interplanetary travel. This is a big advantage of the atomic rocket.

The atomic rocket operates as follows. Liquid hydrogen or some other liquid is fed into a small chamber resembling the combustion chamber of a liquid-fuel rocket. The atomic energy released instantly heats up the hydrogen to an extremely high temperature; the latter is converted into gas and ejected from the chamber under tremendous pressure.

Although the atomic rocket therefore does not differ in principle from the ordinary types of rockets, there are a number of technical difficulties preventing its construction. In the first place the extraordinarily high temperatures and pressures arising in the atomic rocket must be reduced, because no metal can withstand them. Secondly, measures must be taken to protect the astronauts from the radioactive radiations which are released at the same time as atomic energy. To tackle the problem successfully a material must be devised that would absorb such radiations and would, at the same time, be light, because excessive weight would considerably reduce the rocket's range.

Design of Artificial Satellites

The first Soviet artificial moon (Sputnik I) is quite a tiny object: a sphere 58 centimeters in diameter and 83.6 kilograms in weight (Fig. 8.) The satellite body is made of aluminum alloys. Its plating is thoroughly finished and polished, securing good satellite visibility.

High power radio transmitters operating on the frequencies of 20.005 and 40.002 megacycles are installed in the satellite. Its four

Fig. 8 First Soviet artificial Earth satellite (photographed on a stand)

rod antennas are 2.4 to 2.9 meters long. During the rocket's ascent to orbit these antennas are folded close to the satellite plating. But after the satellite separates for independent flight, the antenna rods, attached to it by swivels, unfold and assume the position shown in Fig. 8. The antennas are the only parts on the satellite's outer surface. All instruments with electric power source are sealed inside the spherical body which is airtight and filled with nitrogen. Owing to constant forced circulation the nitrogen regulates the heat exchange between various enclosed instruments and parts.

The second artificial satellite built in the USSR is the last stage of the carrier rocket designed to contain assembled research equipment. Installed in the hermetic satellite body are various instruments for investigating radiation of the Sun in the shortwave, ultraviolet and X-ray bands of the spectrum; for studying cosmic rays, outer space temperature and pressure; an airtight cabin with a test animal (dog), conditioned air system, food supply and devices for the study of vital activities in outer space conditions; a telemetering unit for transmitting the data of scientific measurements to earth; two radio transmitters operating in the same frequencies as those of the first satellite (40.002 and 20.005 Mc), and essential electric power supply sources.

The total weight of the instrument package, test animal and electric power sources is 508.3 kilograms.

Such unmanned artificial earth satellites will, as we shall see below be employed for solving a wide circle of problems that have paramount scientific and practical importance. The dimensions and designs of such satellites will be highly diversified.

In time whole space ship observatories and interplanetary stations will also undoubtedly be established.

Many difficulties still stand in the way of creating a manned satellite space ship. The most complex problem is the need to assure the crew's return to earth, an undertaking of greater difficulty than the launching of a satellite. For this reason alone it is at present still impossible to construct a habitable earth satellite. However the experience that will be gained in the launching and operating of the first unmanned artificial satellites will serve as the basis for the subsequent building of manned satellite space ships.

Although the designs of large satellite space ships are still inadequately developed, it is perfectly clear that their builders will not be constrained by need to streamline their bodies. Since there is no environmental resistance to brake satellite travel, the space vehicle can be of any shape. Thus, for example, a satellite can be given the tore shape (like an auto tire) of a huge circular vehicle with curved tube sections, inside which all conditions will be created necessary for the crew to live and carry on research. Handrails and grappling hooks can be fixed to the structure's outer walls; decks can be installed and the hawsers necessary for space rocket mooring and assembly. The cabins will not be very roomy as satellite designers are obliged to economize weight and, therefore, volume as well, to some extent.

If the tore-shaped vehicle is assembled from separate "cars" coupled together by elastic vestibules, the space ship's size is readily enlarged by inserting new "cars".

The space ship satellite's cabins must of course be hermetically sealed from outer space. Its interior vestibules, doors and partitions must, in addition, provide for sealed atmospheres in separate rooms. Hatches must be clamped shut by locking devices with elastic gaskets. The rubber-framed entry doors, for example, must open inward and not outward. Thanks to this, the pressure of the microatmosphere (the air sealed inside the cabin of a space flight vehicle) will itself help to close still more tightly the cabin doors beyond which is air free empty space. The space ship plating must be whole, seamless. Particularly well sealed must be the contact between metal frames and window panes. These must be of non-shattering safety glass and have approximately the same coefficient of thermal expansion as that of the metal in which the glass is fitted or soldered.

Gas leakage from the space ship cabin does not have much importance on short flights in which spoiled microatmosphere is released anyway into airless space and replaced with new atmosphere. The escape of gas to the outside is, however, extremely undesirable in case the sealed microatmosphere is being continuously regenerated as will be done during prolonged interplanetary flights or in the cabins of permanent satellite stations.

The satellite's glass and metal skin will, as the Earth's atmosphere [does], check the ultraviolet rays of the Sun which permeate [through] universal space and in large quantity are injurious to the human organism. It will also be possible in case of need to draw shut special curtains on the illuminators. Owing to the meteor danger and harmful radiations it may be found impossible in satellite space ship cabins to install windows that open directly to outside space. In this case a narrow passageway with a lens-and-mirror system will have to be installed to bring in light rays. Observations can then be conducted by periscope as in submarines.

Satellite designs have been proposed for a space ship in the form of an air-inflated tore made of rubberized material, glass yarn or impermeable nylon-plastic fabric. But to find a material capable of providing protection from cosmic rays (see p. 56 on harmful radiations) will hardly be possible. For this purpose, the space ship plating will most probably have to be made from various materials in several layers.

In the satellite space ship, as was indicated, the pull of the Earth's gravity is found to be neutralized by the centrifugal force originating in the satellite's circular travel in orbit. People and objects who are on the satellite will therefore be weightless (just as on Earth, due to its headlong rotation around the Sun, we do not feel the force of solar gravitation but only the pull of our planet's gravity). Because of its negligible mass, however, the pull of the satellite's own gravity will be imperceptible.

Satellite space ships can be built having artificial gravity. The satellite is made to rotate around its own axis creating on board a centrifugal force which substitutes for the pull of gravity. A combination satellite can also be built. Weightlessness will reign in one section of it under zero gravity; in another part artificial gravity will be maintained.

The design of a satellite having rotating cabins with intrinsic artificial gravity, is shown in Fig. 9. It is a space station assembled mainly from tanks and other head stage parts of orbital rockets. A satellite space station of such design has built-in properties for constant station extension.

Satellites will not of course suffer a shortage of solar energy. K. E. Tsiolkovsky proposed that the vast streams of sunbeams be captured in outer space hothouses and utilized for growing plants that space island inhabitants might use for food. At the same time the problem of natural air regeneration might thereby be solved. Such a greenhouse would, however, have to be enormous in size. Hothouse area might be reduced, it is true, were designers content to have a single-purpose hothouse to freshen the air but grow inedible plants. In this case also, it would still be a huge structure. Thus, for continuous regeneration of the "portion" of air one man needs, for instance, would require 28 square meters of catalpa bush foliage well-exposed to sunlight. The growing of a special sort of green algae may also prove very useful for restoration of the microatmosphere, as this organism during one hour of sunlight gives off a quantity of oxygen 50 times greater than its own volume.

However, since plants can wither, to rely wholly on the natural air and water cycle is impossible, and therefore an automatic air-regenerating unit on the manned satellite will be necessary.

Early in 1956 the American astronaut Romick proposed the establishment of an artificial satellite in the form of a complete city with 20,000 inhabitants. The Romick project. as far as can be judged from

Fig. 9 Probable design of an artificial satellite. Weightlessness reigns in the lower part of the satellite, but in its upper part artificial gravity has been created by means of rotary movement.

published articles, is based on data repeatedly circulated in astronautical publications and does not contain anything suggesting it might not be practically realized. With respect to the dimensions of a satellite inhabited by 20,000 persons, it must be emphasized that they are not related to the project's feasibility. If a space station can be created for 100 inhabitants, then one can also be established for 1000 and for 20,000 as well. The only issue in doubt is whether or not such a large satellite is practically necessary. Perhaps, in giving the figure of

20,000 for the space island's population, the project's author thus expected to reach more easily the minds and hearts of his countrymen. We Russians are naturally pleased that technical thought continues to find stimulation in the ideas of K. E. Tsiolkovsky, "father of astronautics", on populating outer space with entire "cities" having artificial gravity.

Assembly of Satellite Space Station

Completely new methods of assembly will have to be applied in the construction of large artificial satellites, and after that space stations on the Moon and the planets.

Let us imagine that a rocket vehicle having orbital velocity is large enough alone to accommodate dwellings for people, laboratories, workshops, warehouses, rocketport, etc. That will also be an artificial satellite, a space ship observatory or transfer station for astronauts leaving the earth on a trip to the Moon and the planets. Such a large space station will apparently be constructed as follows. A certain time after a primary rocket has been launched, a second one is sent up and under radio guidance is brought up close to the first. Then in exactly the same manner, a third, fourth and additional rockets will be launched, until a celestial body has been established that is large enough to provide dwelling space for people and accommodate all essential supplies, equipment and instruments.

As to the feasibility of the foregoing, we are reminded that refueling of aircraft in flight has long been practised. This experience can evidently be utilized with success in the construction of artificial satellites, despite the fact that their speed and altitude will be substantially greater. The position and speed of an artificial satellite can be determined at any moment in advance with greater accuracy than is the case for an airplane. In fact an airplane's route and speed depend on the weather and engine operation, while an artificial satellite is quite unaffected by meteorological conditions and travels with the engine shut off.

Delivery by shots from rocket guns has been suggested by Yu. K. Kondratyuk, who writes: "It would be desirable to set up the delivery of the fuel and all objects . . . able without being damaged to withstand accelerations of several thousands of meters a second per second (in appropriate packing—all except delicate instruments) to interplanetary space by rocket-artillery means separately from man . . ."

So, the space station can first be built on earth and tested to the finest detail as to dependability of design and possibility of providing the necessary living conditions for the crew. Then dismantled, the satellite can be dispatched in component parts to the planned orbit where these parts will again be assembled into a whole station.

Naval experience can be utilized in mooring the rockets forwarded from earth to the artificial satellite. From the space station, hawsers (chain cable, for instance) can be thrown to the incoming rocket. The flying units can be pulled together by the mooring ropes.

The following variation is also possible. Instead of the rockets for the satellite under construction being sent in succession, all the rockets are dispatched at once in the form of a squadron. In this case there will be no difficulty in the separate rockets finding those previously launched. The space station can be assembled in a shorter time, since no time is spent waiting the transfer of separate components from the earth. This will in turn lessen the meteor danger for assemblers.

Design of Space Ships

The space ship design will depend largely on its purpose. A rocket designed to land on the Moon will differ in many respects from one intended to fly around it without landing. An Earth-to-Mars space ship will be different from one designed to fly to Venus. There will be a great difference between the rocket using thermochemical fuel and an atomic space ship.

A space ship designed to fly to an artificial satellite of the earth, using thermochemical fuel, will be on the lines of a multi-stage rocket as big as an airship.

Before launching it will weigh several hundred tons, with its payload weighing approximately one-hundredth of this figure. The stages, which will be made to fit onto each other as closely as possible, will be enclosed in a streamlined hull in order to lessen air-drag during flight through the atmosphere. A comparatively small cabin for the crew and a cabin for the rest of the pay-load will probably be designed in the nose. As the crew will stay aboard such a ship only a short time (less than an hour) no intricate equipment will be required.

At the appointed time the rocket will be launched by an automatic starter. Automatic equipment will also be provided to direct the rocket in flight and to take measurements. The "used-up" stages will return to the earth either by parachute or by means of retractable wings converting them into gliders.

The simplest orbital rocket is shown in Fig. 10.

Let us look at still another space ship design (Fig. 11). Such a space ship will take off from an artificial satellite on a journey to the Moon, in order to make a prolonged study of its surface without landing. On completing its journey the space ship will return direct to earth. One can see from the drawing that its principal parts are two coupled rockets, three pairs of cylindrical tanks containing propellant and oxidizer and two space gliders with retractable wings to make a landing on the earth. The space ship will not necessarily have to be streamlined, because it will be launched from a site beyond the upper layers of the atmosphere. See pages 48-49 for Fig. 11.

A space ship of this kind will be built and tested on the earth, following which it will be taken apart and shipped to an interplanetary station. Fuel, equipment, food and oxygen will be sent there separately.

Once the space ship is reassembled at the interplanetary station it will start on its journey into space.

The propellant and oxidizer will be fed into the engine from the central cylindrical tanks, which in fact will be the space ship's principal chambers, filled with fuel needed in the first stages of the trip. The crew will suffer the inconvenience of staying in the cabin of the glider until the principal chambers are emptied, i.e., for a few minutes after the rocket is launched.

The remaining fuel will evaporate immediately when a small valve, connecting the tanks with the empty space, is opened. Then air will be pumped into the tanks and the astronauts will travel in them to the end of their journey.

At a certain distance from the Moon the space ship will become a satellite of that body, using the propellant and oxidizer of the lateral tanks in the stern for the purpose. When the fuel is used up the tanks will be jettisoned.

The astronauts will not switch on the engines again until the time arrives for their return journey, for which the fuel will be provided from the lateral tanks in the nose. Before entering the earth's atmosphere the crew will move back into the space glider, which will then be detached from the main body of the space ship circling the earth.

Fig. 10 ONE OF THE POSSIBLE DESIGNS OF AN ORBITAL ROCKET

I Head stage
II Middle stage
III Tail Stage

  1—artificial satellite
  2—powder blocks
  3—combustion chamber
  4—nozzle
  5—fuel tank
  6—oxidizer tank
  7—compressed gas tank for feeding fuel to combustion chamber
  8—fuel pump
  9—oxidizer pump
  10—turbine
  11—fuel pipeline
  12—oxidizer pipeline
  13—swivel mounting of engine
  14—automatic pilot
  15—automatic pilot lever for guiding engine

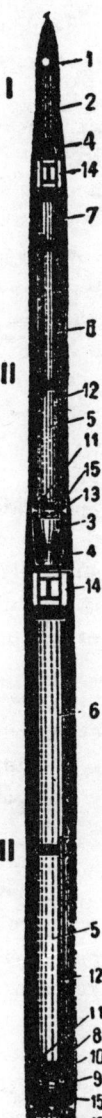

To make a safe landing the astronauts will make use of the retractable wings of the glider.

The space ship will have much in common with a submarine in that its crew will be obliged to live in an airtight cabin, completely isolated from the external medium. The composition, pressure, temperature and humidity of the air inside the rocket will be controlled by special apparatus. But there will be one big advantage where the space ship is concerned. The difference between external and internal pressure in the cabins will be less for the space ship, there being only one atmosphere (the external pressure is equal to zero), while for the submarine, there are frequently several atmospheres of pressure.

Of great importance is the problem of maintaining the necessary pressure in the space ship cabin. It is possible that with constituent elements in the air selected in a certain manner, the astronauts can breathe normally when pressure in the cabin is below one atmosphere. Meanwhile, the less the pressure [is] in the space ship cabin, the thinner can be the walls of its hull, the simpler it will be to construct the cabins and space-suits and the less danger of the leakage of air into empty space in case of insufficiently sealed skin connections or a puncture formed in the skin from a meteorite hit.

In the Earth's atmosphere oxygen deficiency is usually felt at a pressure of 430 mm of mercury column, which corresponds to an elevation of 4.5 kilometers above sea level. Experiments have established that when the pressure of inhaled air is lowered, its oxygen content should be increased, otherwise asthma can result.

When environmental gas pressure is below 47 mm of mercury column the human organism, just as any organism, cannot exist even in pure oxygen since at such pressure (corresponding to atmospheric pressure at 19 kilometers altitude) the water contained in the human body (at 37° C) begins to boil. Such an occurence is, of course, dangerous for life: cracks appear on the skin, cells rupture, caisson disease or "the bends" result. At low pressure the functioning of the hearing organ also deteriorates and the teeth ache. Meanwhile a rise of pressure is just what is needed to reduce evaporation from the body surface. Choice of the most expedient pressure can be determined only by experiment.

The microatmosphere in the ship cabin must not consist of pure or almost pure oxygen.

Experiments have established that at a pressure of 190 mm of mercury column, pure oxygen causes the same physiological reactions as air does at sea level. However with prolonged breathing pure oxygen has a debilitating effect on the organism.

When oxygen is present in large quantity, it is noteworthy that the fire hazard is intensified and food products rapidly turn sour and spoil. The microatmosphere must therefore contain other gases as well. In this connection the possibility has been repeatedly discussed of having some kind of inert gas to replace the nitrogen in the microatmosphere of a space ship or artificial satellite.

The air in the cabin can be continuously purified by cooling it in a special condenser to the liquefaction temperature of carbon dioxide, that is, to minus 78°C. Whereupon water will first precipitate, and liquid carbon dioxide also. Oxygen and water vapors must be added to the purified air in necessary quantity; the microatmosphere's water content has therefore to be reduced. The mixed atmosphere must then be heated to normal temperature.

To forward a heavy air-regenerator to an artificial satellite designed for short term service will hardly prove expedient. It will obviously be simpler to replace the spoiled microatmosphere with a fresh one by means of "airing" the cabin. Chemical means can also be used for removal of carbon dioxide from the microatmosphere. Thus in a submerged submarine, for example, absorbents are employed successfully to remove carbon dioxide during many days of underwater travel.

The necessary supply of oxygen can be conveniently carried in liquid form. Oxygen can be shipped from the earth to the space station even in solid form; in which case the container can be very light. The oxygen for respiration can likewise be stored in the form of a sodium peroxide compound which absorbs carbon dioxide gas and excess moisture, and gives off oxygen. Hydrogen peroxide in solid form would be a still more suitable way to carry oxygen supplies.

Gravity being absent in the weightless conditions of the space station, its microatmosphere is without the natural phenomenon of convection or air exchange between lower and upper layers. Unless corrected this situation can cause the air to stagnate with formation of carbon dioxide "sacks" making impossible both the process of respiration and that of combustion, thereby asphyxiating the space ship occupants. The air in the cabins must therefore be constantly stirred by fans or other means, and forced unilateral air circulation simultaneously maintained.

CHAPTER TWO

Man in Outer Space

Safety of the Organism at Cosmic Speed

Let us now examine the problem of whether or not man can endure the physiological strains of flight in outer space, and especially, of ascent to a satellite space ship, living on it and subsequently descending again to the Earth's surface.

Indisposition during such a space voyage can originate chiefly in disturbance of normal sensations of gravity or weight. First of all we note that unless its acceleration is excessive there is no speed beyond human endurance. As a matter of fact, are we in the least troubled by the Earth's rotation around its axis? Yet, due to the Earth's rotation, points on the Equator attain a travelling speed of 1675 kilometers per hour. Are we disturbed by the Earth's travel around the Sun at a velocity exceeding 100,000 kilometers per hour? Do we, finally, observe the movement of our whole solar system through outer space at the rate of 70,000 kilometers per hour? With these facts in mind, we can assert that the human organism is able safely to endure any speed.

A round-trip flight from Earth to outer space and back again can be compared to a giant leap up in the air during which astronauts will at times be affected by the intensified pull of gravity, and at others experience absence of gravity or weightlessness. Similar effects are observed during an ordinary high or broad jump. When with a hop, we jump off the ground, we feel our body's increased weight. This stage of the jump is similar to the rocket take-off from the Earth's surface. The moment our feet leave the ground and our body is hurtled across a space of air, we fly by our own momentum, or coast, without feeling our weight. This stage of the jump is comparable to a rocket ship coasting with engine turned off. When our feet finally touch Earth again, the braking of our speed begins, and we again feel our weight. This third stage of the jump is similar to the breaking period in the descent to Earth.

In the World of Intensified Gravity

When a train jerks to a start the passenger feels a push from behind and he is jolted against the back of his seat. What is called overload or G strain occurs, its main cause being accelerated motion. Overload affects the organism just like gravity, only more intensely. The G strain of overload, caused by the rocket engine pull, is precisely what the astronaut in a rocket cabin will also experience. In take-off this pull is naturally greater than the clutch of terrestrial gravitation, otherwise the rocket would not rise from its launching pad. Whence comes the name "overload" (3 G, 5 G and other multiple G strains are spoken of, the G standing for the gravitational force common to earth).

When an aircraft takes off by catapult, the flyer endures a 4 G strain; that is, he experiences four times his usual weight. During aerobatic figure or stunt flights pilots frequently endure 8 G strain, and for aquatic sportsmen diving into water a 16 G strain is ordinary during the plunge. It is, however, necessary to bear in mind that the G strain on a catapult lasts but a few seconds, and when diving into a pool, or more precisely when braking in water after the plunge, it is endured mere fractions of a second. Moreover, in ordinary means of transportation the gain in speed in gradual when acceleration is low. However, the examples cited do not by any means prove that man can for a sufficiently prolonged period endure the G strain necessary to reach circular velocity in an orbital space ship.

Before space travel is realized can it not be established what G strain man can endure without danger to life and for how long a time? Yes.

The centrifugal force originating in rotary motion also generates G strain. G strains of any magnitude and long duration are thus obtainable. To test human endurance a man is placed in the cabin of a turntable machine, a kind of whirligig, which is set rotating at variable speed. With appropriate selection of turning radius and rotation speed, exactly the same sensation can be aroused in a test subject as would be the case in a given rocket during take-off. Results obtained by such experiments show that a rocket crew can endure the acceleration (and G strains) attendant on development of astronautical speeds. Most people can safely endure for some minutes 4 to 5 G strains.

The degree of human endurance under G overload depends to a very large extent on the body position while the rocket engine is running. When in a sitting position, for example, the human organism reacts differently to G strain than it does in a prone position on the back or face downwards. The standing man feels gravity most of all in his feet. But in other positions the body feeling of weight and also general body fatigue will differ. Thus, we tire less sitting than standing, and least of all when lying down. A most effective way to reduce fatigue during G strain is to have the person put on a special tight-fitting jacket (a proposal the author advanced in 1933 in a work submitted to the Committee on Astronautics in Paris). Experiments have shown that special "counter-gravity" suits which provide intensified pressure in the legs and lower parts of the trunk to retard the outflow of blood from the head and facilitate blood supply to the brain, will enable a human being easily to endure 3 G strain for an interval greater than the engine running time of an orbital rocket.

Capacity to endure high G strains, we note, depends partly on the individual organism characteristics and partly on training. Strains which one person endures with comparative ease may prove fatal to another.

Life in Zero Gravity Conditions (Weightlessness)

When the rocket ship is coasting through empty cosmic space on its own momentum, the people on board feel no weight. And this is why: the sensation of weight derives from the pressure of a support (a floor, chair, bed, etc.) under the body and the reciprocal pressure some body parts exercise on others. If the support be removed, however, the sensation of weight vanishes as well.

Let us explain this in an example.

If we place three bricks one on top of the other, the uppermost brick will exercise a certain pressure on the one in the middle, while the latter's pressure on the lowest brick will be twice as great. If we throw the three bricks out of the window, however, they will in falling no longer press upon each other, none of them being a support for either of the others.

In the literature on astronautics the term "weight" is generally taken to mean a force that keeps the people and equipment on the floor of a space ship cabin, and not the Earth's gravity pull, which, it stands to reason, never disappears. Astronauts will feel just the action of this pressing force and not the pull of the Earth's gravitation, which is what puts tension in the springs of a scales (spring balance), in the string of a plumb line. In the absence of this gravity force people and objects do not exercise any pressure on each other and become weightless.

The Human Organism at Zero Gravity

The effect of weightlessness on man has been studied in part with success during high altitude flights of jet aircraft. In conducting experiments the plane climbs to a high altitude, and at the moment of climb reaching maximum speed, the engine is cut out. The plane then coasts in very rarefied layers of atmosphere like a cast stone, experiencing only negligible air-drag. In these conditions the force of gravity almost completely disappears ("almost," since a certain atmospheric resistance nevertheless remains). In a jet fighter it is thus possible to create a condition of weightlessness for about one minute. Moreover, the effect weightlessness (although incomplete) has on people and animals can also be studied during the power dive of an airplane or the sky-diving of a parachutist in a delayed drop.

Experiments have shown zero gravity sensation of one-minute duration to be harmless for man, although in the first moments he loses any control over his movements, which become very jerky.

At the VII International Astronautical Congress S. J. Gerathewohl (USA) made a report on the results of 300 experiments made to examine the effect of temporary absence of gravity on 16 men during special test flights carried out in 1955-1956.

Sensitivity to the physiological effect of weightlessness was found to be extremely varied, not only in different people but even in the same person, depending on circumstances. While many people perceived the loss of gravity with satisfaction, others became more or less seriously ill. Some experienced nausea, the effects of which were felt for many hours after landing. During temporary weightlessness, pathological processes were not observed in blood circulation, and voluntary head movements in the absence of gravity did not cause unpleasant sensations. The influence of zero gravity depends a lot on training and how long the stay in the weightless state lasts. It is very risky to apply the results of zero gravity's transient effect in terms of its effect over a prolonged period of time.

The sensations people have when in a weightless state have been found to be extremely subjective, as was stated above. Thus, a 35-year-old pilot who had behind him a thousand hours of jet flying time and had been subjected more than 200 times to the action of weightlessness, mostly in a jet plane he himself piloted, reports that movement of his extremities was effortless and muscular coordination was not at all disturbed. To orient the plane with reference to the Earth's surface is not difficult. The "up" and "down" directions are not confused. The weightless state aroused a pleasant sensation in the pilot. With the absence of gravity he did not observe any kind of adverse symptoms with respect to vision, hearing and respiration.

Gerathewohl himself, after having experienced the weightless effect for 47 seconds, described his sensation as follows: "Never in my life did I feel so good, and were I free to choose my kind of rest, my choice would certainly fall on the state of weightlessness."

Another test subject, 46 years of age, who had a wealth of experience flying gliders, reports on the contrary that at zero gravity he lost his sense of "up" and "down." Among those test subjects who fell ill from zero gravity effects were found both 20-year-old novices and flyers aged 30 and over who had 1000 to 1500 flying hours behind them.

Mice, dogs and monkeys endured the state of weightlessness—incomplete, it is true—for three to four minutes, considerably longer than humans. In experiments conducted in the USSR under A. V. Pokrovsky's direction, dogs ascended to an altitude reaching 110 kilometers. During the rocket's fall back to Earth, the dogs in spacesuits were, with an appropriate device, catapulted out into space: one at an altitude of 90 to 85 kilometers, and another at 50 to 35 kilometers altitude, and from the altitude of 4 kilometers down descended by parachutes to the ground. All nine test animals landed without mishap (three dogs did this twice). The motion picture films, electrocardiograms, temperature, pulse and other measurements made on the dogs during the flight showed that in the main animals can adjust very well to the absence of gravity, with varied reactions, depending on individual characteristics.

A temporary weightless state is thus seen not to be harmful for the human organism. In any case it is harmless for many people. The stay in an artificial satellite or space ship can, however, last many days or even whole weeks and months, and therefore we can as yet construct only more or less well-founded conjectures about the state of health of future astronauts. Some investigators think that the heart will function normally also when zero gravity is prolonged, inasmuch as heart activity resembles the mechanical action of a closed cycle pump. The heart has to overcome only the friction of blood on the walls of vessels. It is however impossible to rest on such reasonings since cardiac activity is closely connected with the central nervous system.

Breathing problems in the weightless state are more complex. For example, during a short fall (in particular, during a parachute jump), a catch in breathing is usually observed. Flight in an artificial satellite will, owing to absence of gravity, be sensed by man just as a fall is, and if the flight lasts a long time, it will perhaps be necessary to employ devices for artificial respiration.

Food can be eaten even when gravity is absent, since the passage of food depends on muscular contraction of the esophagus (liquid foods can be taken even when the head is lower than the body).

Ordinarily the physiological functions proceed under any body position—standing, sitting or lying. Consequently, a change in the position of body organs with reference to the direction of gravitational force has no substantial effect on the way they function. Certainly, it is very difficult to hold the head lower than the body for any length of time. This shows that in certain unusual body positions the force of gravity has a harmful effect on the organism. But this by no means indicates that for other normal body positions the presence of the force of gravity is essential. Quite the contrary, on the grounds that most of the physiological functions are completed under the action of muscular forces, osmotic processes (infiltration through porous membranes or partitions) etc., we have every reason to hope that the absence of gravity will not introduce substantial derangement in the organism's functional activity.

Such are the results of investigation into the effects weightlessness has on the living organism in terrestrial conditions. The launching of the second Soviet satellite with a test animal on board opened up completely new possibilities for study of how adaptable organisms are to a world without weight. In Sputnik II the action of prolonged weightlessness on the living organism was for the first time subjected to experimental examination. This experiment proved to be encouraging: the physiological behaviour of the dog which travelled on the satellite in a specially-equipped airtight cabin, was satisfactory not only for just the first few hours but for several days.

This gives us grounds to assume that the astronaut will not, when experiencing loss of gravity, lose self-control even at the outset of a space flight and will during that time be able to create on board the flying space vehicle tolerable living conditions (see page 51).

Working and Living Conditions in Absence of Gravity

We will now dwell on physical aspects of everyday life in conditions of weightlessness on a satellite space ship, living conditions which will naturally differ substantially from those we are accustomed to on Earth.

At zero gravity the normal perception of "up" and "down" disappears. An object dropped from the hand will not fall and "down" will become a conditional direction toward the center of the earth. People can rest in any body posture. Walking is rendered impossible since essential foot pressure against a support will be lacking, and consequently the friction necessary for locomotion will also be missing. Movement from place to place inside the artificial satellite or interplanetary ship will be possible by pulling and pushing oneself toward and away from the walls or other fixed objects.

Bodies will be weightless on an artificial satellite, but it should be observed that to bring them into motion or, vice versa, to stop or slow them down, a definite force has to be applied to them for an interval of time since they naturally do not lose the property of inertia inherent to any mass.

In case an astronaut goes out from the satellite vehicle into empty space, he will obviously have to remain connected to it by means of a rope. He can take with him a heavy object attached to a rope, and by throwing the object in one direction, move himself in the opposite direction (utilizing the natural law of conservation in position of a mass's center under the action of internal forces alone). The same effect can be achieved by means of a small rocket or pistol, but these methods involve irretrievable loss of the mass.

It will not be possible to use ordinary furniture or tools. Any object wanted at a definite place will have to be fixed there. Flasks with liquid, for example, will have to be anchored to the wall. When cooking food the pots will have to be covered with lids and set in rotary motion by means of a special centrifuge so that their contents adhere to vessel walls. Electromagnetic utensils, which should operate well at zero gravity, will be very convenient to use.

When liquid is poured from a vessel it is, through surface tension, converted into a sphere or globule. When it comes in contact with a solid body, its cohesive force can exceed the force of surface tension; then the liquid will flow out over the body surface. Handling of liquids will in general be quite inconvenient. One can only wash himself with a wet sponge or towel. To empty a bottle it will literally have "to be pulled" off from its liquid contents or swung neck-out in a wide arc to utilize centrifugal effect, or siphoned with a pump or rubber syringe bulb.

If at zero gravity one strikes a match on a box, the head will flare but the match will not light. Neither will gas ignite or a candle burn. The reason for this is that under normal terrestrial gravitation the combustion products, the hot gases, being lighter, ascend upward by natural convection and give access to a fresh supply of oxygen necessary to maintain the flames. But in the absence of gravity the gases, not being lighter than the air around them, accumulate near the flame and extinguish it. To maintain flames a continuous stream of oxygen has to be fed the burner. Electrical heating utensils will of course be more convenient in this respect.

In weightless conditions the dust in the air will not settle on the floor and other surfaces, which makes it a health hazard. To combat dust, electrofilters can be employed. The astronaut's clothing must be so made that it remains on the body regardless of the force of gravity. Many things and functions will thus be somewhat difficult to manage on a space ship. But certain activities, such as shifting loads from place

Fig. 11 (Center) Space ship designed for round-the-moon trip. I—Taking off from an artificial satellite of the Earth; II—Space ship becoming an artificial satellite of the Moon; III—Path around the

Moon; IV—Space ship heading away from the Moon; V— Separation of the gliders when the space ship approaches the Earth; VI—The gliders land on the Earth.

to place, for example, will be substantially eased in absence of weight.

Let us make a final important remark. Mention is sometimes made of the "apparent" increase of weight or the "apparent" loss of weight in a flying rocket. Such a view is profoundly erroneous. Increase and loss of weight are quite real phenomena and they can be determined by means of instruments.

Fig. 12 shows how the weight of a body changes in the course of

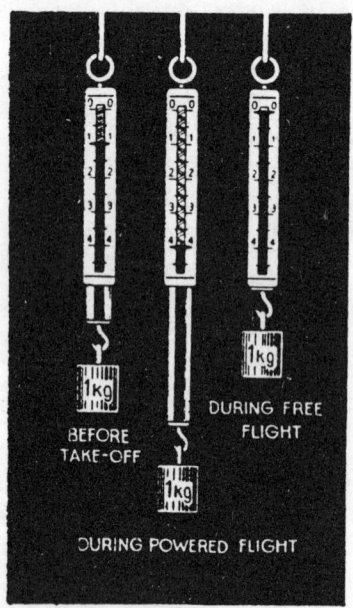

Fig. 12 Variations in the weight of a body during interplanetary flight

space flight. On earth before take-off a one-kilogram weight suspended from a spring-balance moves the pointer to the one-kilogram mark. As soon as the rocket is up in the air, however, the weight of the bodies inside increases several-fold, for instance, four times as much, and the pointer of the spring-balance points to the four-kilogram mark. But during the flight on its own momentum all objects in the space ship lose weight and therefore the pointer of the spring-balance passes to the zero mark.

Artificial Gravity

Experimental proofs that man will feel perfectly normal in conditions of weightlessness have not yet been forthcoming. Special medical preparations may possibly have to be used to maintain normal functioning of the human organism.

Simplest at first glance appears to be the creation of an artificial field of gravitation sustained by continuous engine operation although at lowered power, as was proposed by Esnault-Pelterie (1912). This method would, however, involve excessively large fuel consumption. Meanwhile, an extremely simple method for creating artificial gravity exists; namely, rotation of the artificial satellite. According to an idea which K. E. Tsiolkovsky advanced as early as the end of the last century (1895), the space vehicle (for example, the ship pictured in Fig. 11) must be made up of two interlocking sections. At the necessary moment these sections are separated from each other in space but remain connected by cables; then by means of small rocket motors the sections are made to rotate around their common center of gravity (Fig. 13). In the airless medium of empty space, once the system has reached the requisite angular velocity its further rotation will obviously continue without propulsion.

To sum up, we see that from the viewpoint of physiology life on a manned artificial satellite or space ship presents no insurmountable obstacles. The astronauts can no doubt endure 4 G strains for some minutes while the rocket engine is running. This will permit communicating to the rocket the necessary speed at sufficiently economical engine operating conditions.

With respect to flight in a coasting rocket after the engine has been cut out and flight in an artificial satellite, we are not altogether certain that the absence of gravity for a prolonged period of time will be harmless to the human organism. But even should it be harmful, this must not become an obstacle in the way of creating manned artificial satellites and space ships, since it is wholly possible technically, by means of rotation, to create the sensation of gravity.

The reassuring experiment of the test animal's flight on board the second Soviet satellite suggests that after the engine stops working, the space flyers will not lose self-control and will apparently be in condition to set up the artificial gravity system. In any case, they

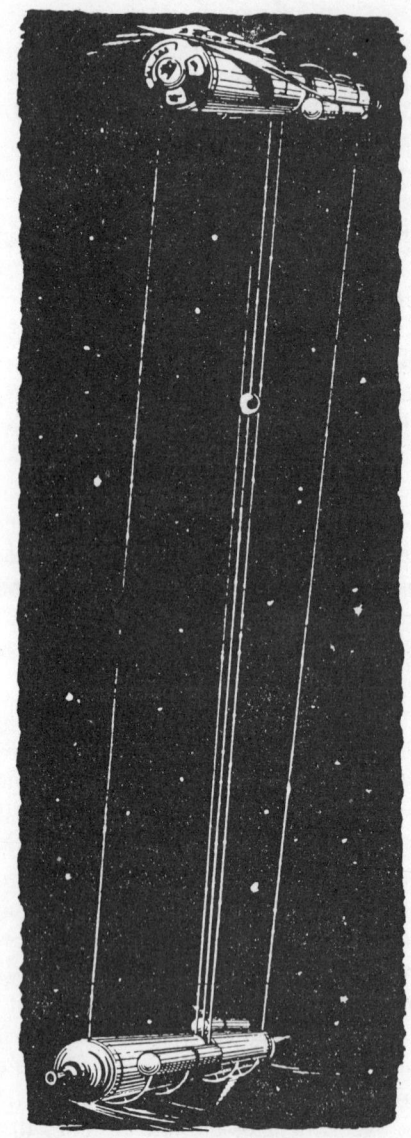

Fig. 13 Creation of artificial gravity on the space ship

can passively await the moment when a automatic control puts the satellite or space ship into gravity-creating rotation.

Food and Respiration Problems

The problems of providing astronauts with oxygen, water and food in the hermetically sealed cabin of a flying space vehicle are already solvable, however much work still remains to be done by experts in refinement of solutions. And these problems are not secondary, if it be taken into account, for instance, that an interplanetary expedition can last for several years. Very little work has yet been done also in examining the possibilities of air conditioning and water regeneration on board the space vehicle.

A wide difference of opinion exists among various authors concerning life-sustaining rations of food and respiration for astronauts. Some figure this ration below 4 kilograms, others as much as 10 kilograms, a day per man. The dry food ration varies from 0.5 to 1.2 kilograms. The lower limit seems closer to the truth for perfectly dry, dehydrated foods. A little more than one kilogram of oxygen is enough for oxidation of these products (it should not be forgotten that the respiratory process is closely connected with the digestive: the more we eat, the more oxygen we also absorb). Since carbohydrates in combination with oxygen yield fewer calories than proteins and fats, the latter will have to predominate in the food of astronauts.

In high-grade food products, however, the carbohydrates in combination with oxygen can successfully compete with fats in heating power. Thanks to this the menu of astronauts can be diversified with a minimal weight of provision stocks. The favorite food of North American Indians, pemmican, a paste of dried meat, fat and berry juices, will probably be highly valued on a satellite space ship because of its high calorific value and capacity for long preservation.

The water ration per man-day is set at 2 kilograms. The suggestion that water and oxygen be taken along in the form of hydrogen peroxide seems reasonable. This will allow for reduction in the volume of the oxygen tanks, since the oxygen is, as it were, packed in the water. Moreover, when the hydrogen peroxide decomposes into water and oxygen, a certain amount of heat is given off, which can be used for heating dwelling quarters.

An installation for regeneration of water can be found profitable only for interplanetary ships and for permanent satellites. Besides, the total amount of water on board will be increasing all the time on account of the synthetic water which is made constantly in the organism during oxidation of the hydrogen contained in the dried food products.

Dangers of Space Flight

The fear is frequently voiced that a space ship runs the risk of collision with large meteoric bodies. How great is this danger and what are the means of protection from it?

The Earth is subjected to a continuous bombardment by meteors, and in the course of a year several thousand meteors land on the surface of our planet. Meteors are iron or stony bodies of various sizes, some of them being several meters in diameter before entering the Earth's atmosphere. As far as meteorite particles are concerned, these fall to the Earth at a rate of between ten and one hundred thousand per second. The total weight of all meteorite bodies reaching the Earth's surface in a day is estimated at ten to twenty tons, their velocities outside the Earth's atmosphere being between ten and seventy kilometers per second.

Meteors are heated in the atmosphere as a result of friction against the air and at times become as bright as the Sun or even brighter. A crater several kilometers in diameter may be formed following the crash of a meteor onto the Earth's surface. A space ship could be destroyed if it were hit by a meteor of appreciable size and even a minute puncture in the hull would be dangerous, causing the air to leak out at the speed of sound. Experiments have shown, however, that a man would not lose consciousness for about fifteen seconds after the sudden drop in external pressure, which is quite sufficient to enable him to put on the oxygen mask of a space suit.

Microscopic meteorites might also destroy the ship's skin if they bombarded it long enough. This is particularly dangerous for artificial satellites revolving around the Earth over prolonged periods of time. "Constant dripping wears away a stone," as the old saying goes.

In the experiment conducted in the U.S.A. in 1953 at heights between 40 and 140 kilometers 66 hits were recorded during 144 seconds, or 4.9 hits per second per square meter. From other experiments we learn that polished metal plates projected to a great height had tiny dents—traces of hits by micrometeorites—when they were examined with a microscope on their return.

Effective methods of protecting the space ship from the meteorite menace have not yet been worked out. However, some progress has been made in this field. We know, for example, that meteors are not uniformly distributed in time and space. A number of meteorite showers and the time of their fall-out have been studied, and there has been detailed research into the orbits of the numerous swarms of meteors. The resulting information will help astronauts to choose a correct trajectory and start on their journey at the right moment. During a "meteorite calm" they will be able to reach the Moon and return home without running the risk of encountering any sizable meteors on their way. We were convinced of this by the experience of the first artificial satellites which travelled a distance many times greater than such a trip without any "meteoric" wreck. The outer metal plating of the space ship will provide reliable protection from meteorite dust, while the inner skin will shield it against small meteors.

When the space ship gets beyond the Martian orbit it will face another danger—that of colliding with one of the smaller planets or asteroids revolving around the Sun, chiefly in the space between the orbits of Mars and Jupiter. The astronomers have spotted and charted the routes of about 1,600 such planets.

The total mass of the smaller planets is approximately the same as that of all the meteorite matter of the solar system (about one-thousandth of the Earth's mass). It is perfectly clear that a collision with any of these bodies, the smallest being one kilometer in diameter, would be the end of the space ship.

To avoid collision, radar equipment can be used to give a timely warning and to automatically divert the rocket from its route. This is a difficult problem, however, because of the enormous velocities at which meteorite bodies travel in space.

Autopiloting radar gear that diverts the space vehicle's path in flight may possibly find successful use to counter larger meteorites. Owing to their enormous cosmic speeds, space ships, it should be observed, nevertheless lack the "maneuverability" which surface means of locomotion have. For a pedestrian to step around an obstacle in his path is very easy; for the auto driver roaring along at high speed to detour is more difficult, while for the rocket plane it is still harder to turn rapidly out of the way of approaching craft. To deflect a space ship rapidly from its orbit is a feat that is naturally most difficult of realization. Much research still faces scientific thought in this field.

Another possible way to protect against the meteor danger successfully is to shoot the meteoric bodies out of the way. The shooting must, it stands to reason, be automatic from radar detection of menacing meteors to range-finding and firing of anti-meteor machine guns. The meteoric body "sighted" by radar coming out of space toward the satellite or space ship, is blown up some distance off after being hit by the shell, and only a minute portion of the "spatter" can land on the flying vehicle. These fragments will apparently be no more dangerous than meteoric dust, from which the ship's outer plating offers protection enough.

An artificial satellite will be somewhat easier to protect from meteoric dust than an interplanetary ship will be. The upper layers of the atmosphere can, in fact, also provide partial protection from meteors for a satellite if the vehicle travels low enough over the earth's surface, orbiting at an altitude such that the air is, on the one hand, already so rarefied that it offers almost no resistance to satellite movement, and on the other, is dense enough for security against the "speediest" small meteors. Such an altitude can be assumed to be approximately 200 kilometers. While air density at this height is tens of millions of times less than it is at the Earth's surface, the small meteors do not generally penetrate to this level of the atmosphere.

In passing through "astral rains," when meteoric bodies fall in showers on the earth, artificial satellites will possibly have to submerge into denser layers of the atmosphere (but not below 100 kilometers), which will also serve them as partial protection. To overcome the air-drag that will then develop, it will be necessary to switch on a small rocket motor, and also take measures to protect the satellite plating from overheating.

Harmful Radiation

For all life on Earth our planet's atmosphere is known to be a reliable protection from harmful radiations of the Sun, which are absorbed by the oxygen in upper atmospheric layers, during its conversion into ozone.

The ozone is located in the stratosphere at variable altitudes from

16 to 50 kilometers. Consequently, even an artificial satellite flying quite low will not be protected from the action of the Sun's ultraviolet rays. It is true that recently high altitude rockets have successfully established that the ultraviolet radiation of the Sun is considerably less intensive than had been surmised on the basis of observations made from the earth. Therefore, the plating of the flying space vehicle itself will possibly be found sufficient protection from these radiations. In case this is not so, the gap between the skin layers can be filled with a layer of oxygen which acting similarly to atmospheric oxygen, after being converted to ozone, will create a barrier against the ultraviolet rays of the Sun.

Artificial satellite or space ship illuminators will also play a part, since even ordinary window glass is known to absorb most ultraviolet rays. These rays in small quantity are, we note, essential for the normal functioning of the organism (hygienists think that house windows should be made of glass that admits ultraviolet rays).

Injurious X-rays are also present in the solar spectrum, but it is not difficult to counter them on flying space vehicles, since they are easily absorbed by almost all building materials.

Outer space is likewise permeated by cosmic rays, most of which are blocked by the earth's atmosphere. These rays consist chiefly of protons and alpha-particles (nuclei of hydrogen and helium), and travel at a velocity close to that of light. If it be considered that every ten minutes, a million, perhaps even several millions, of such fast missiles will bombard every square meter of space ship surface, it then becomes clear what a serious danger astronauts confront in cosmic rays. But the danger for the human organism is not confined to the so-called primary radiation. When cosmic rays collide with the skin of the ship or space-suit, what is called secondary radiation will result just as it does when cosmic rays strike the earth's atmosphere.

The effect comic rays have on the human organism is as yet very little known. The laboratory experiments in this field are only in a rudimentary stage. Of great interest from this viewpoint are the experiments carried out in the United States with small animals which were taken up by stratostat to an altitude of 30 kilometers for about 30 hours. The exposure of black mice to primary cosmic rays resulted in the appearance of gray spots considerably larger than expected on their skin. In other experiments no diseases of any kind were observed even after a 24-hour exposure. The test animals must remain for a prolonged period after the experiments under careful observation.

To ascertain the effect of cosmic rays on the human body the following test has been carried out by the Swiss scientist Eugster. A small piece of preserved human skin was carried to the upper layers of the atmosphere in a high-altitude rocket and exposed to cosmic rays. After the rocket's return the skin was grafted onto a man, and the graft took. Consequently, it had not lost viability. Similar experiments were then conducted on a wider scale in the United States. High-altitude rocket research has proved that brief exposure to both ultraviolet and cosmic rays is harmless not only to the lower animals but to monkeys, too.

But how will cosmic rays affect people who remain on artificial satellites or a space ship for a long time? This question remains so far still open.

Lead armor plating of artificial satellites might conceivably be a way to protect the crew from primary and secondary cosmic rays. A layer of lead 2 to 4 centimeters thick blocks 35 per cent of the cosmic rays at altitudes above 100 kilometers. But inconceivable difficulties attend the transfer of even a substantially thinner armor to an artificial satellite.

The lives of the travellers in an atomic rocket would also be endangered by radioactive emissions from nuclear fuel. Some parts of an orbital rocket will possibly acquire artificial radioactivity, which can have a harmful effect on the living organism for a long time after the rocket engine stops working. It is therefore essential to develop special, very light, protective shields against these harmful radiations.

Training for Space Flight

The minutes of flight when the engine is switched on will be the most tense for the crew of a space rocket. Success in orbiting an artificial satellite depends largely on achieving the necessary speed at a certain altitude and a precise travel direction at that moment. The chauffeur, the helmsman, or the pilot can always correct any deviation of an automobile, ship or airplane from an indicated path. But the guidance of a space rocket, in distinction from surface craft, is a very complicated business in the take-off not only because the pilot experiences intensified gravity, but also because he must react instantly.

Quite naturally, therefore, astronauts will have to have appropriate physical training. They will have to possess an organism of heightened resistance to the strains of an overload and an absence of gravity, to the low barometric pressure of the microatmosphere in the space ship, to large fluctuations of temperature. The piloting of a space ship and other operations requires not only profound craftsmanship, but also great skill and serious preparatory training. The biological factor, the man, his health and hardiness, will therefore play not the least role in accomplishment of space flight.

When preparing for flight in a space rocket, the crew can be trained in a rotary machine which can be set to revolve at a rate such that its centrifugal force will accelerate just as the pull of a space ship's rocket engine does under real conditions. To provide training in the endurance of prolonged weightlessness is, however, quite impossible under conditions on Earth.

The prolonged training of space men, and also the manifold testing of various designs of space vehicles in laboratory "dry runs" under simulated flight conditions, is a mandatory stage enroute toward realization of manned orbital and interplanetary rockets, and satellite space stations.

Such training can be carried out in what is called a prototype or simulator of a space ship cabin. Thus, in the opinion of Amico (USA), for example, the space ship cabin simulator must be equipped with scientific measuring devices, control instruments and other units, and have life-saving gear as well. When all the devices are set in motion, the physical conditions of a space flight are precisely simulated. The mock-up cabin must be an exact duplicate of the cabin in a projected space rocket or artificial satellite. Simulated in it should be air pressure, temperature, acceleration, lighting, various radiations and so forth. The conditional travel speeds, the vehicle's declivity in relation to the sky, the pressure in the tanks and pipelines, the fuel consumption, etc., must be under automatic control.

During ideal flight at top speed, navigational problems connected with travel in a provisional trajectory will also be solved. The testing unit must also include a system of remote control by radio.

In simulating the active segment of the trajectory in which the engine runs, account must be taken of reduction in the force of gravity and air density with altitude, of variation in aerodynamic resistance, change in the rocket's mass, particularly, with the successive jettisoning of auxiliary stages.

Any air pressure can be simulated in a barochamber, jet propulsion action can be replaced by centrifugal force, etc., but it is impossible, for instance, to create zero gravity under laboratory conditions. The simulator will no doubt suffer from other defects as well, since it is difficult to mock-up the various dangers that will menace astronauts in flight, such as cosmic radiation, collision with meteoric bodies, etc., and these shortcomings can nullify the whole experiment.

When performing duties during the "take-off," the spacemen must wear protective "counter-gravity" suits (see page 42) to test the possibilities of working in such conditions. The crew must not only master the normal routine of servicing the space rocket's equipment but must also be able to make use of life-saving devices at top speed. (Training in life-saving methods is extremely important. In aviation, for example, the schooling of airmen includes a training program in coping with all possible accidents that might occur but once in the course of 20 years' service.)

Life-saving gear should be tested for effectiveness in use in case of a puncture caused by a meteor hit, in case of intense heating or deep cooling of the space ship plating, in case the oxygen unit, the guidance or signal system or the electric network fails to operate, etc.

Spacemen in flight must constantly check the good working order of automatic controls and so must learn to follow instrument readings and react to them appropriately.

What astronauts do in the cabin simulator must in turn be controlled and recorded from the outside. The mistakes they make can then be analyzed.

Under the direction of control instructors the crew must at first master separate operations like navigation and life-saving, and only after elaborate training will the trainees make a general test of a complex "cosmic vehicle"—for example, an orbital rocket or artificial satellite. At this time the actions of the crew in the simulator cabin will no longer be controlled by individual instructors, but by a staff of specialists in various branches of science.

Such a research program is necessary not only to achieve the final purpose—rational design of space vehicles and training of skilled personnel to operate them—but is required also for daily progress in the spaceflight field. The training of jet aircraft test pilots is, we note as an example, usually of several years duration.

CHAPTER THREE

Artificial Satellite Orbits and Observations

Artificial Satellite Travel

An artificial satellite cannot fly above the Earth over any route as an airplane does. There is not the least possibility of placing an artificial satellite in orbit around tropical parallels or around polar circles. A satellite cannot possibly be made to fly along a zigzag line. To shorten or lengthen considerably the time a satellite takes to fly from one city to another is impossible.

An artificial satellite can fly only in a circular or elliptical orbit. Like a body thrown at an angle toward the horizon, the satellite can moreover travel only in a plane passing through the Earth's center (Fig. 14), that is, in the plane of a great circle. The artificial satellite cannot, therefore, in particular, travel above any parallel of the terrestrial globe, the only exception being the equator, which is zero parallel. The plane of an artificial satellite's orbit will remain fixed in relation to the arch of the sky.

Sidereal Time of Artificial Satellite's Rotation

The flight altitude of an artificial satellite depends on its travelling speed and the corresponding duration of its period of circuit around the Earth.

Were there no air resistance, an artificial satellite launched from the earth's surface with the above-indicated speed of 7912 meters per second would complete a full circuit in the sky around the Earth, returning to a previous position with reference to the stars and the earth's center in one hour, 24 minutes and 25 seconds. This is what is called the sidereal circuit period.

When its launching altitude is higher, the artificial satellite's orbit grows longer and the pull of terrestrial gravitation becomes weaker.

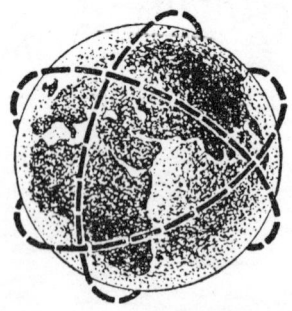

Fig. 14 An artificial satellite can travel only in a plane passing through the Earth's center

Consequently, the centrifugal force can also be less and the satellite's travel slower.

The circuit period of a satellite increases with recession from the planet. At an altitude equal to two radii of the Earth, the sidereal circuit is 7 hours, 17 minutes, and at altitudes two to three times greater still it is 15 hours, 44 minutes and one day, 2 hours, 3 minutes respectively.

The satellite's circuit period can be calculated as follows. Knowing the flight altitude of the satellite and the radius of the globe, we determine the length of the circular orbit; that is, the path over which the satellite passes during one circuit of the globe, and then we divide the result obtained by the circular speed. For example, the radius of the orbit of a satellite in flight at an altitude of 6,378 kilometers is equal to 12,756 kilometers, and the length of the corresponding circumference is 80,152 kilometers. After dividing this magnitude by the circular speed, equal to 5.595 kilometers a second, we derive 14,327 seconds or 3 hours, 58 minues and 47 seconds.

The Satellite Circuit Period Relative to an Observer

The star or sidereal circuit period of a satellite flying in orbit near the Earth's surface is, as was indicated above, one hour, 24 minutes and 25 seconds. Such is the satellite circuit period with reference to the sky overhead or to an observer who is at one of the Earth's poles. But let us imagine that the satellite's orbit is in the plane of the equator and the satellite is, just as the Earth, circling from west to east. By the time the satellite makes a complete circuit of the globe with reference to the sky, the observer at the equator has, together with the Earth, been swung around at a rather large angle in relation to the sky, and owing to this he is found to be a great distance ahead of the satellite. Only after 5 minutes and 16 seconds does the satellite catch up to the observer.

With reference to the observer, the circuit time of the so-called zero artificial satellite is thus one hour, 29 minutes and 41 seconds. This is the time that elapses until the satellite returns to its previous position in relation to the observer. In other words, the observer again sees the artificial satellite in the sky in its former position with respect to himself, for instance, in the zenith.

Were the satellite to travel in the plane of the equator along a circular orbit from east to west, then the observer at the equator would, as it were, travel toward it. For a satellite travelling from east to west at the earth's surface, the circuit period with reference to an observer would be 4 minutes and 41 seconds shorter than the sidereal circuit time and would amount to one hour, 19 minutes and 44 seconds.

Orbits of the First and Second Artificial Satellites of the USSR

The first Soviet artificial earth satellite travels in an orbit inclined at an angle of 65 degrees to the plane of the equator. The maximum height of its flight reaches 900 kilometers.

The satellite orbit is an ellipse with a slight contraction: the difference in length between its major and minor axes amounts to less than a quarter of a per cent. This orbit is evidently almost a circle. However, the center of this "circle" is somewhat displaced with reference to the center of the Earth.

The satellite's circuit period was at first 96 minutes and 12 seconds, and then owing to atmospheric resistance (although extremely slight), began to shorten. During the first three weeks the circuit period shortened approximately 2.3 seconds a day and after 23 days amounted to 95 minutes and 18 seconds. Thus, the number of times the satellite circled the globe was, during that interval, increased from 14.97 to 15.11 circuits per day .

Orbit dimensions are also reduced with shortening of the circuit period. Calculations show that during the first 18 days the major axis was cut down approximately by 70 kilometers (a half per cent).

The plane of the satellite orbit, without changing its inclination in the plane of the equator, is slowly turning around the axis of the Earth. During one satellite circuit, this motion is roughly one quarter of a degree in longitude, in a direction counter to the Earth's rotation.

From the data of official reports, the average travelling speed of the satellite during the first three weeks of its existence, is readily calculated to have been 7.58 kilometers a second. The satellite's orbital travelling speed is however not constant. It is somewhat greater in the northern hemisphere, where the satellite flies lowest, and is somewhat less in the southern hemisphere, where its flight altitude attains maximum height.

The rocket carrier which had placed the satellite in its orbit, was at first left behind it, because the impulse with which the satellite was detached from the carrier rocket, somewhat slowed down the flight of the rocket itself. The carrier rocket was in addition braked more intensely by the air than was the satellite itself. In four days after the satellite was launched, the carrier rocket lagged behind it only one thousand kilometers. Then the carrier overtook the satellite and even went ahead of it. This occurred for the following reasons: The decline in the rocket's orbital travelling speed brought about a shortening in the length of its orbit, and the carrier rocket's circuit period became shorter than that of the satellite. The angular velocity of the rocket's travel through the sky thus became greater than that of the satellite, and for the observer on earth, the carrier rocket therefore outdistanced the satellite.

From the moment when the carrier rocket passed the satellite (from October 10, 1957), it began to travel more to the east of the satellite in longitude, that is, began to intersect a given parallel more eastward than the satellite, and its angle of intersection from day to day became increasingly larger. The question therefore arose in many minds as to whether or not these two bodies were rotating in different planes, intersecting at an increasingly greater angle. But in fact both orbits lay in one and the same plane. The carrier rocket's passage more eastward than the satellite's is, however, explained by the Earth's rotation and the carrier rocket's outrunning the satellite. On October 24, 1957, at 1800 hours, for example, the carrier rocket thus intersected a definite parallel and went on travelling all the time in a plane that was fixed with reference to the stars. Travelling behind it, the satellite reached this same parallel one hour later. But during that hour the earth had turned westward around its own axis a distance of 360° divided by 24 hours, which equals 15°. Consequently, the satellite intersected the mentioned parallel 15° farther westward than the point where the carrier rocket had intersected the parallel. In other words, the carrier rocket passed over 15° eastward of the satellite in longitude, although both artificial celestial bodies were orbiting in one and the same plane.

Compared with the first artificial satellite, the second one is travelling in a higher orbit, which at maximum distance from the Earth's surface is about 1700 kilometers out in space—that is, 800 kilometers higher than the first orbit's apogee. The period of one complete satellite circuit amounts to one hour, 43.7 minutes. This is 7.5 minutes longer than the first satellite's circuit period at the start of its travel. The orbit's angle of inclination to the plane of the equator is approximately the same as that of the first satellite (65°).

On the basis of the laws of celestial mechanics, it is readily calculated that the perigee of the second satelite passes over at approximately the same altitude as did the perigee of the first satellite at the beginning of its existence. Inasmuch as the major axis of the satellite orbit amounts to approximately 14,700 kilometers, and the circuit period equals 6222 seconds, the satellite's average orbital speed is then derived as equal to 7.41 kilometers a second.

The second satellite will probably hold out in cosmic space longer than the first and its circuit period will be only negligibly reduced. It now ascends to heights such that air friction is practically no longer encountered. Were the satellite's entire orbit to be at an altitude of 1700 kilometers, its life duration would be figured in years. Only on the orbital segment adjoining the perigee does the satellite encounter any kind of air-drag, this too being negligible. Nevertheless, under the influence of this factor, the satellite speed will in time gradually decline.

Satellite Space Station

Since celestial bodies exercise on each other a mutual gravitational attraction, it is impossible to construct an Earth satellite which would remain motionless in outer space. Such a satellite is out of the question. But it is possible to create a satellite which, while travelling around the Earth with reference to the stars, will nevertheless be fixed in the sky with reference to an observer on Earth.

As a matter of fact, the satellite's period of circuit around the Earth is, as was already pointed out above, increased with increase of the satellite's distance from the Earth. A satellite which is orbiting at an altitude of 265 kilometers requires an hour and a half for one circuit around the Earth, whereas the Moon which is almost 400,000 kilometers distant from the Earth, requires about four weeks to circle the Earth. There is also obviously a distance at which the satellite would complete one circuit in exactly one day.

If such a satellite is, moreover, travelling in the plane of the equator, and from west to east besides, then its angular velocity will be equal to the angular velocity of the Earth's rotation around its own axis, and it will thus prove to be motionless as seen by an observer on Earth. Such a satellite—let us call it an artificial space station—must be at an altitude of 35,800 kilometers above the equator, as can be calculated. The attraction of the Moon, it is true, will cause a certain derangement of the satellite's orbit that will with time disturb its "immobility," but these disturbances can be eliminated by appropriate correction of the trajectory.

So as to picture better the possibility of creating a "fixed" space station, let us assume that a tower 35,800 kilometers high is erected at the equator. In proportion to ascent up this tower, centrifugal force gradually increases (because of the increase in the radius of rotation around the Earth's axis), whereas, on the contrary, the Earth's gravitational pull diminishes. At the very top of the tower both these forces counter-balance each other, or are put in equilibrium. Now, imagine a gondola is attached to the tower top. From the foregoing it is clear that if the tower now be removed, the gondola will not fall down, but remain where it is fixed at this distance from the Earth and rotating

Artificial Satellite Orbits and Observations

together with the Earth. To an observer on the Earth, the gondola will seem stationary; it will have become an artificial satellite space station.

A satellite space station will have a number of advantages over other Earth satellites. In fact, from such a satellite our whole planet would seem motionless, its apparent diameter appearing approximately 40 times greater than the Moon's diameter seems as seen from earth, and the area of its visible disc 1600 times larger than the Moon's is. A satellite space station's crew would more readily communicate with earth by means of directional radio waves or light signals. A flight to the satellite station could be made at any time without waiting on the necessary alignment of the satellite with respect to the launching pad.

Such countries as Indonesia, Brazil, Colombia and others that lie on the Equator could construct satellites hanging "stationary" over the territory of the country or even "swinging" over it (in the case when the orbital plane forms a small angle with the plane of the equator). But were it necessary to construct a space station observatory for inspection of one of the European countries, or the whole of Europe, then such a satellite will certainly have to fly over other countries and continents as well.

Artificial Satellite Observations

The visibility of an artificial satellite depends not only on its size, reflective capacity, distance to it and so forth, but also on the light contrast with reference to the sky background. Such a "star" can therefore be observed only at dawn and evening twilight when the satellite is lit by solar rays, and darkness has descended on the earth surface where observations are made. Water vapors and dust suspended in the atmosphere greatly impair satellite visibility. But the main difficulty is the briefness of its period of visibility. Twilight can last longer than the satellite's circuit period with respect to an observer. Therefore, the satellite might possibly be seen twice during one twilight; that is, after the satellite has been observed to set, its next rise in the sky may be observed during the same twilight.

An artificial satellite's surface might be covered with a phosphorescent substance to facilitate observations. Illuminating the satellite inside will achieve the same purpose.

It is noteworthy that in case a satellite's radio transmitter gets out of order, visual and radar observations remain the only means of determining its orbital travel.

Owing to the high flight altitude of a satellite, its travel can be traced from a substantial part of the Earth's surface, and enormous expanses of the globe will be visible from it.

The first artificial satellites are easily recognized among other heavenly bodies shining in the sky, since in distinction from them, the satellites travel across the sky in northern, northeastern, southern and southeastern directions, and not westward or to the north and southwest, as the Moon and stars do. The explanation of this is that the first artificial satellites were launched in the same direction as the Earth's rotation and not against it.

In the course of time, when more powerful rockets come into use, it will also be possible to launch satellites in a direction contrary to the Earth's rotation, in which case the satellite will be indistinguishable from other heavenly bodies in the apparent direction of its travel through the sky.

To conduct observations of artificial satellites from various points on the globe, there is no need to launch many satellites in different directions. For example, it is enough that a satellite appear once over the North Pole for it then naturally to fly over the South Pole as well. Such a satellite's orbit will continue always to pass over the poles. Let us imagine an artificial satellite, circumnavigating the globe 16 times a day at an altitude of 287 kilometers above the poles (265 kilometers above the equator). From such a height the Earth looks to the observer like a huge disc occupying a large part of the sky. What appears like a "cap" of our planet 3700 kilometers in diameter will be visible from the satellite. This "cap" will be shifting all the time. During one satellite circuit, the Earth will make one sixteenth of a rotation on its own axis, and the "cap" will be shifted at the equator by a distance of 40,000 divided by 16, or 2500 kilometers. During 16 satellite circuits in a single day the whole terrestrial globe will float before the orbiting observer's eyes both in daylight and under cover of night.

For the observer on Earth, an artificial satellite orbiting at an altitude of 200 kilometers is, at the moment it passes the zenith, travelling in the sky with just the same angular (apparent) velocity as an airplane flying at 7130 meters altitude with a speed of one thousand kilometers an hour, or as a plane flying at 3065 meters altitude with a speed of 500 kilometers an hour. Therefore, to keep a moving artificial satellite in the field of vision, is obviously not difficult. The apparent travelling speed of artificial satellites in orbit above 200 kilometers will be still less. And it is precisely at high altitudes that artificial satellites will, as a rule, fly in the future.

Should the artificial satellite not be at the zenith, but lower in the sky, closer to the horizon, then the apparent speed of its travel across the sky will be less, and it will be easier to track. At the moment of rising (or setting) in the sky the satellite will, with respect to the observer, be travelling with the least speed. But in proportion to the satellite's ascent in the sky, its angular velocity relatively to the observer (the apparent speed of the satellite's travel in the sky sphere) will rapidly grow, the speed acceleration being faster the lower the satellite orbit lies (Fig. 15). If, for example, the orbital altitude of an equatorial satellite is equal to the radius of the earth, then the speed of its angular travel with reference to the observer will, at the zenith, be double what it is at the horizon. But for an equatorial satellite, travelling at 300 kilometers altitude, the speed of its angular travel at the moment it is observed in the zenith will be already 22 times greater than its angular velocity with respect to an observer at the horizon. After passing through the zenith overhead, the artificial satellite begins to slow down its movement with reference to the observer, and at the moment of setting its angular velocity falls to the magnitude it had when rising.

The higher the satellite's flight altitude is, the greater the area is of the Earth's surface from which it will be visible (Fig. 16). Thus, a satellite flying, for example, at an altitude of 200 kilometers, will be visible from a territory with a radius of 1500 kilometers. But with a flight altitude of 1000 kilometers, the visibility range is doubled.

The higher the satellite's flight altitude is, the longer will be the interval in which it can be observed from a definite point on Earth. Thus, for example, a satellite flying at an altitude of 200 kilometers will be visible for seven minutes, at 500 kilometers altitude for 11 minutes, and at 2000 kilometers altitude for 28.5 minutes.

Can an artificial satellite that has once been observed suddenly appear later over some unexpected strip of territory? No, it cannot. Satellite travel is strictly confined to its orbit. If a satellite's coordinates be determined at a certain moment and its speed and travel direction at that moment be recorded, it is sufficient to establish by calculation for any moment of time in advance, the satellite's location and to forecast when and what territories it is destined to fly over. In particular, it can be ascertained if the satellite is to appear again

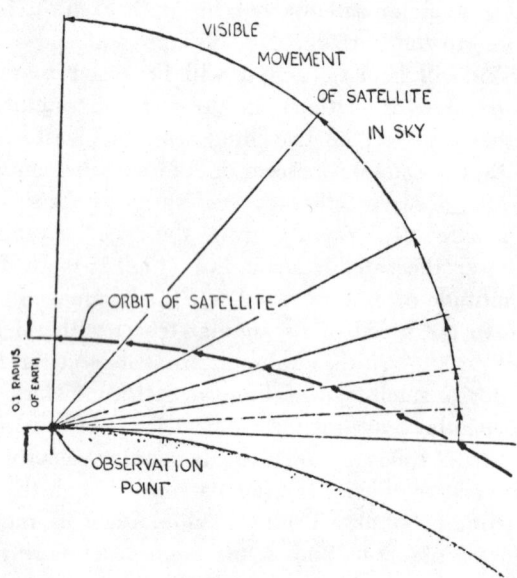

Fig. 15 At the moment of rising the satellite will, from the observer's viewpoint, travel at the lowest speed. But as it rises in the sky, the satellite's angular velocity with respect to the observer—the speed of its apparent travel across the sky—will in proportion increase rapidly. After passing the zenith above the observer, the satellite begins to slow down its movement relative to him, and at the moment of setting its angular velocity falls to the magnitude which it had on rising.

over a given locality and, if so, precisely when. That is, of course, assuming that the satellite's direction and travelling speed are not changed by rocket engine action.

The same artificial satellite can fly over a given locality once from south to north, then again from north to south, not because the satellite has suddenly changed its travel direction in orbit (which is impossible), but because the Earth, during the time intervening between

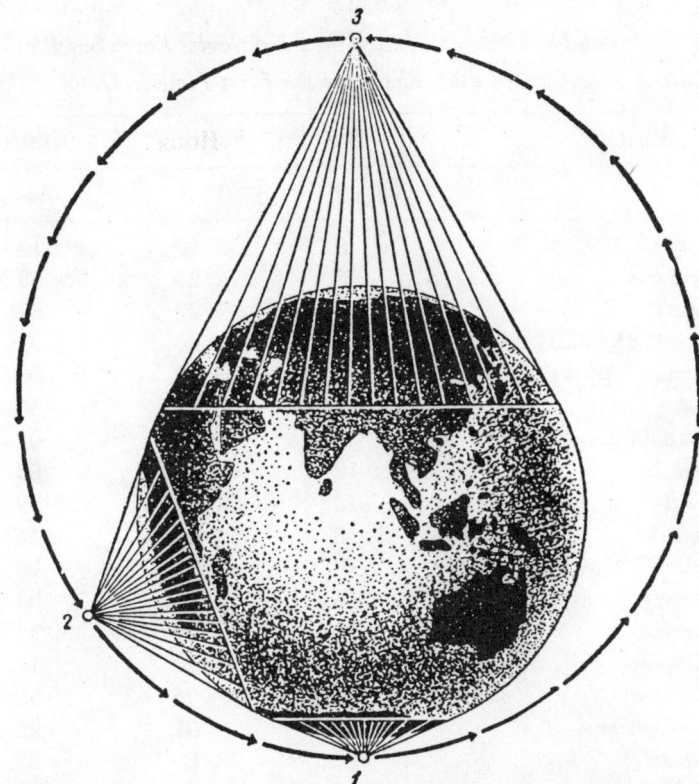

Fig. 16 Enlargement in diameter of the globe's visible segment with altitude of Earth satellite's flight. From an altitude of 500 kilometers the diameter of the visible global segment amounts to 4900 kilometers (1); from an altitude of 2000 kilometers to 9000 kilometers (2); and from an altitude of 7000 kilometers it increases to 13,700 kilometers (3).

one observation and the other, has rotated halfway around its axis. Thus the satellite's passage through the sky is reversed in direction for the observer.

Observations of the First and Second Artificial Satellites

Owing to the orbits of the first two artificial satellites being inclined at a large angle to the plane of the equator, they flew over almost all

TABLE I.

Places and the Dates on which the First Soviet Earth Satellite (Sputnik I) Appeared over Them for the First Time in October, 1957

Place	Day	Hour	Minute
1	2	3	4
Algiers	9	10	14
Alma-Ata	5	20	58
Ankara	10	22	37
Antarctic (Coast)	7	06	—
Askinus (Alaska)	10	18	03
Athens	12	22	36
Ashkhabad	8	21	02
Baku	12	21	00
Beirut	13	06	59
Belgrade	8	00	14
Berlin	13	00	16
Bombay	6	07	03
Brussels	10	08	34
Budapest	10	00	18
Bukharest	13	06	55
Buenos-Aires	12	01	25
Warsaw	12	06	52
Washington	5	16	31
Vil'nius	10	06	57
Damascus	6	08	34
Delhi	8	19	24
Jakarta	8	16	01
Dublin	13	08	29
Erevan	6	22	36
Kabul	6	20	58
Cairo	8	22	35
Calcutta	5	19	16
Karachi	5	20	54
Kiev	7	00	15
Copenhagen	8	01	53

TABLE I. (*Contd.*)

Place	Day	Hour	Minute
1	2	3	4
Leningrad	6	06	49
Lisbon	11	10	09
London	6	10	05
Madrid	9	10	12
Melbourne	8	12	38
Mexico	6	18	16
Minsk	10	00	19
Moscow	5	01	46
New York	7	06	36
Oslo	6	03	27
Ottawa	8	14	57
Paris	6	10	06
Peking	7	17	49
Prague	6	01	49
Pyongyang	11	16	10
Reykjavik	12	05	07
Riga	6	01	51
Rome	6	10	09
Rio de Janero	8	15	18
Istanbul	11	22	35
Thallin	8	06	59
Tashkent	7	21	01
Teheran	8	07	01
Tirana	8	00	13
Tokyo	6	16	11
Ulan-Bator	5	19	23
Frunze	6	21	01
Hanoi	7	17	43
Helsinki	8	01	56
Shanghai	10	16	10

continents and water expanses of the globe (except for the polar regions and narrow belts situated south of the Arctic circle and north of the Antarctic circle). Their travel thus covered almost 90 per cent of our planet's surface.

The passage of the first Soviet artificial Earth satellite was registered in all corners of the globe (Table I). Sputnik I was observed by the naked eye as a star of the fifth to sixth magnitude, and its carrier rocket as a star of the first magnitude.

Shown in Fig. 17 is the trace left on a photographic plate by the

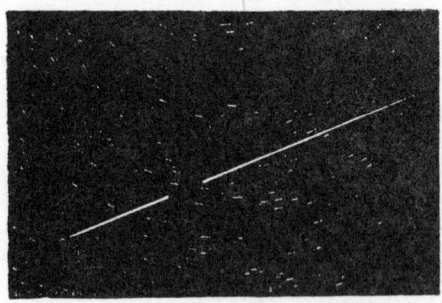

Fig. 17 Trace left by carrier rocket on photographic plate during prolonged exposure (drawn from photograph by T. P. Kiseleva, Central Astronomical Observatory of Academy of Sciences USSR [Pulkovo], October 10, 1957). The photo shows that the carrier rocket (compare length of tracings) moves through the sky at an apparent speed several times faster than the daily travel rate of stars. The gap in its track indicates the time and duration of the carrier rocket's passage through the sky.

carrier rocket (of Sputnik I) during a long exposure. It is evident in the photograph how much faster the carrier rocket travels across the sky than the stars do in their daily apparent movements. The relative velocities are proportional to the length of the tracings left on the photographic plate (the astrograph used to take the picture remained stationary). It is also evident in the photograph that the carrier rocket travels at an angle to the direction of the daily star movement. The gap in the carrier rocket track indicates the time and duration of its passage across the sky.

The first artificial satellite travelled, as already mentioned, in an ellipse, very nearly a circle. Due to the Earth's rotation, a projection

Artificial Satellite Orbits and Observations 75

of the satellite's orbit onto the surface of the globe forms an extremely complex curve. Graphed in Fig. 18 is a projection of satellite travel onto the surface of our planet during a period somewhat longer than one of its circuits around the globe. After one complete circuit, the satellite is at the zenith, not above the place it started from, but above another point on the same parallel situated approximately 24 degrees farther west than the first. (Were the displacement exactly equal to 24 degrees, then one day later, the satellite might be observed from the same place and in the same position. But in reality there are slight deviations.) Plotted in Fig. 19 is a diagram of the satellite's travel during one 24-hour day. Between territories around which the satellite flies there remain, as is evident, belts outside the range of the orbital travel—areas where the satellite never appears overhead in the zenith but where it can be observed at a certain angle to the horizon. Should the satellite's life be sufficiently prolonged, however, it can nevertheless appear in the zenith above these places also. In the very first days of the satellite's existence, the global points of its projection were observed to shift somewhat in comparison with preceding days. By the end of November, 1957, Sputnik I had already completed enough circuits of the Earth to have streaked the whole map of the globe had its trajectory been graphed thereon.

Immediately after the first artificial earth satellites were placed in orbit, their travel was tracked in the Soviet Union by 66 visual watch stations and 26 radio observation stations. In addition, observations were made by radar, radio direction-finding and other lookout instruments. The radio tracking done by many radio amateurs in various corners of the globe is naturally the most prevalent form of satellite observation.

The two radio transmitters on board the first and second satellites emitted radio waves of 7.6 and 15 meters in length. This was a great convenience for radio amateurs, many of whom do not have receivers that can operate on the very short waves. The satellite radio transmitters emitted signals in the form of telegraph sendings about 0.3 seconds long with a pause of the same length. The signals of one frequency were sent during the pause in the other frequency signal.

Sputnik I's radio "beep" signals were received at distances reaching several thousands of kilometers, and in exceptional cases, ten to fifteen thousand kilometers.

After three weeks of continuous radio transmission, Sputnik I's

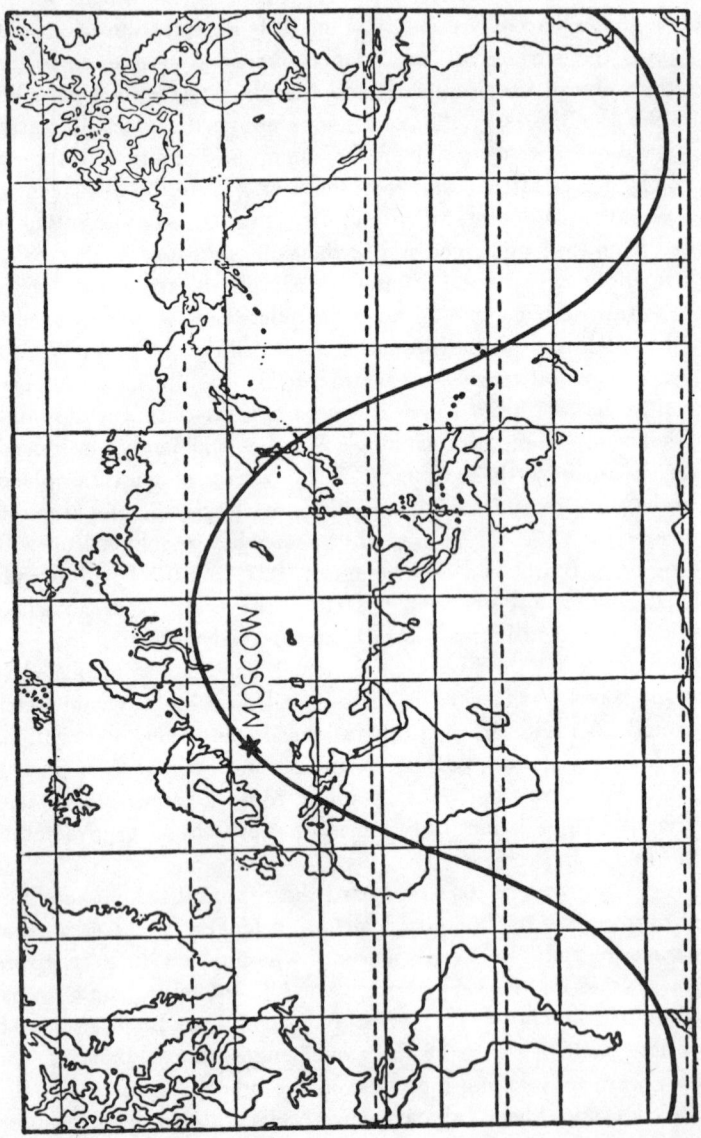

Fig. 18 Diagram of first artificial satellite's travel during slightly more than one of its trips around the Earth

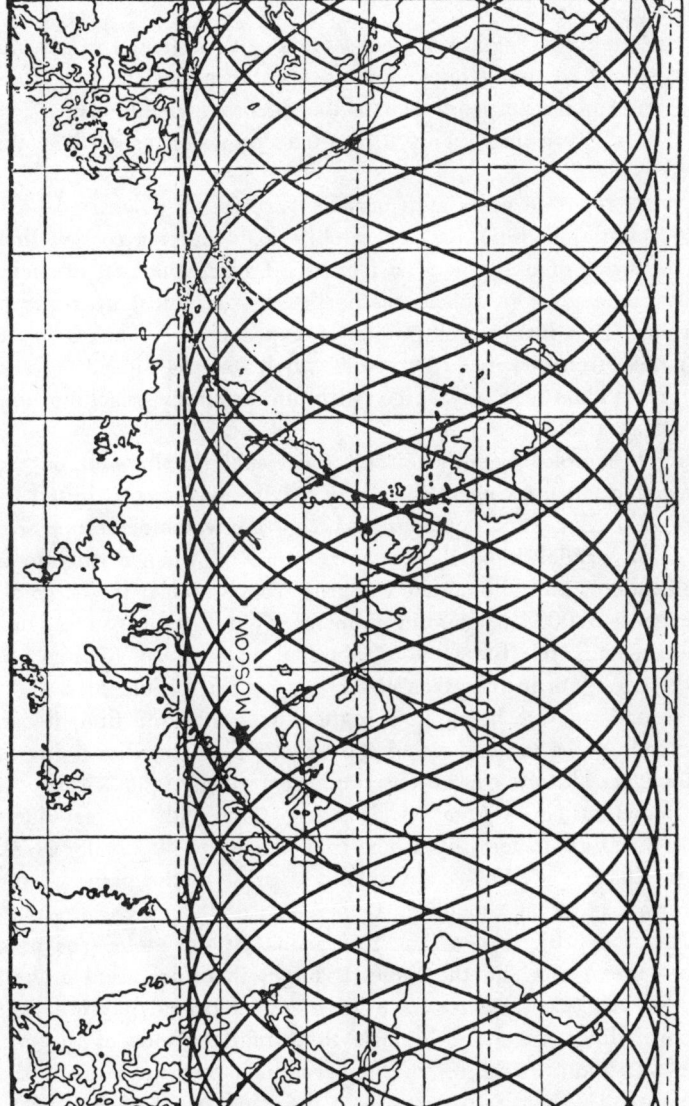

Fig. 19 Diagram of first artificial satellite's travel during one day

electric power supply was exhausted. Subsequent observations of this first satellite and its still brighter carrier rocket were made chiefly by visual methods.

For the satellite visual watch, observers with optical instruments were organized in two groups: one to observe along the meridian, the other in a plane perpendicular to the satellite's visible orbit. Two optical "bars" were thus set up in the field of vision crossed by the satellite.

The press reported daily ephemerides (previously calculated positions of celestial bodies) for the satellites and carrier rocket, thus easing the work of observers. To facilitate finding the first artificial heavenly bodies ever to appear in the sky, astronomical instruments were aimed in advance at "points of expectation." The error in angular measurements made by visual watch stations did not exceed one degree. Time was measured with an accuracy reaching one hundredth of a second.

During a complete satellite circuit the visual watch point on the earth's surface was, owing to the Earth's daily movement, shifted by 1100 to 2600 kilometers, depending whether the watching station was on the 65th parallel or on the equator. However, when a satellite is at an altitude of 200 kilometers, the area from which it is visible already exceeds 3000 kilometers in diameter. Therefore, in roughly one hour and a half after the satellite sets at a given place, it often appeared rising again in the given observer's field of vision.

The second satellite being in a higher orbit than the first, it can be observed over a larger area of the Earth. If the satellite flying at 900 kilometers altitude can be observed theoretically from a range of 3200 kilometers (in a great circle arc); then with the satellite's ascent to 1700 kilometers this range is increased to 4200 kilometers.

The inclination of the second satellite's orbit to the plane of the equator was, as already mentioned, the same as that of the first satellite. And the projection of the two orbits on the globe was at a definite moment one and the same. It might therefore seem at first glance that the two satellites have an identical travel diagram. But their travel diagrams differ because the circuit periods of the two satellites vary, and therefore rotating around its own axis inside the satellite's orbits, the Earth occupies a differing position relative to each of them.

The second satellite completes 13.9 trips around the Earth in a 24-

hour day. If it be assumed, therefore, that the plane of the satellite orbit remains stationary with respect to the stars, then after each circuit the satellite would intersect a parallel of the globe shifted 25.9 degrees to westward. However, as we know, the plane of the satellite orbit is slowly turning with respect to the stars. Accordingly, the distance between points of the satellite's routine intersection of the Earth's parallel amounts to roughly 26.3 degrees.

Star Movements Seen from Earth Satellites

No sooner does the orbital rocket, a minute or two after take-off, fly out beyond the limits of any perceptible atmosphere, than the sky loses its usual blue color and becomes pitch black. Absolute darkness does not descend on earth even in places shaded from the Sun because, dispersed in one degree or another by the atmosphere, the solar light penetrates the zone of shade. But in contrast to this, almost total darkness reigns in cosmic regions plunged in the shadow of any non-luminous celestial body. The sky is not luminous there with bright sunshine dispersed by atmosphere, as there is none. Only the feeble light of stars and nebulae gleams. Stars do not twinkle but are distinctly visible all the time, when the eyes are screened from the direct solar rays. Otherwise, having become adjusted to the Sun's bright light, the eyes lose their capacity to distinguish the stars.

From an artificial satellite the view of the heavens will differ very much from what it is on the surface of Earth. The greater part of the sky over the southern hemisphere is inaccessible to our eyes, the eyes, that is, of people living in the northern hemisphere; likewise the major part of the northern sky is out of visual range for inhabitants of the southern hemisphere. From an artificial satellite, irrespective of its travel direction, the whole sphere of sky can be observed during indigenous "days" (that is, during one circuit around the Earth). To an observer on a satellite, it seems that during an indigenous day the Earth has made a complete circuit around the satellite.

If the satellite orbit is circular, then the Earth's movement seen in the satellite sky sphere will be uniform. But in case of an elliptical orbit, owing to the satellite's own travel irregularity and also because distance to the Earth is variable, it will seem to the observer on the satellite that the Earth at one time accelerates, at another slows down its movement across the sky.

We have described the apparent movement of celestial bodies as it would be on an artificial satellite in orbit around the Earth. Let us now imagine that we are on board an Earth satellite having artificial gravity generated by satellite rotation. In this case what movement do we see in the satellite sky? It will seem to us, first of all, that the sky together with the Earth, Moon, Sun and stars is circling around the satellite. One such circuit in the sky sphere will be completed in the time interval it takes the Earth satellite to make one turn around its own axis of rotation, i.e. a few minutes or even fractions of a minute.

If the satellite rotation axis is horizontal, it will seem to the astronauts that the Earth first turns over their heads, then dips down underfoot. But if the satellite's rotation axis coincides at a certain moment with our planet's axis of rotation (since an artificial satellite cannot "stand still" over the pole, such a coincidence lasts an instant), it will then seem that the Earth rotates around its axis with incredible speed, hundreds of times faster than it really does, and that the Sun and stars circle the Earth at the same fantastic speed. Depending on the satellite's rotation direction, this apparent Earth rotation can appear to be forward or backward.

Were the satellite's axis of rotation (which must, as we know, pass through its own center of mass) to pass nevertheless through the center of our planet without coinciding with its axis, the Earth would then seem to be rotating not around its own real axis, but around the satellite's rotation axis. To the astronauts the Earth's pole will seem to be that point on its surface from which the artificial satellite will be visible at the zenith. For space travellers moving around the Earth, therefore, this imaginary terrestrial pole will seem to be migratory, but capable of slight travel during one apparent circuit of the earth around its "axis."

In a particular case the apparent pole of the Earth's rotation can also appear to be stationary. This will happen in case the Earth is observed from a satellite space station rotating around its own axis passing through the satellite's mass center and the Earth's center. Such a pole, for example, can turn out to be the Kenya mountain range in equatorial Africa with its banana plantations and coffee trees. From the viewpoint of observers in space, the Kola Peninsula lying beyond the Arctic circle and Sumatra situated on the Equator will in this case be on one apparent parallel.

As we see, the astronauts in artificial satellites will be obliged to exercise much ingenuity to master the art of cosmic navigation and utilize practical astronomy, for instance, in correcting the satellite orbit.

Data on the movements of celestial bodies in relation to unmanned satellites can also be useful for interpreting results of the measurements of instruments installed on these space ship observatories.

Day-Night Cycle and Annual Seasons on Earth Satellites

A strange new world will open before mankind when, following in the trail blazed by unmanned space explorer satellites, people are sent into outer space and find themselves inhabitants of a new heavenly body, a manned Earth satellite. The sky will seem quite unusual. Annual changes of season will differ and other things will seem out of place.

Day and night will alternate on the artificial satellite as they do on Earth; but then the days won't be the same as earthly days. Since an artificial satellite can in one 24-hour day circle as many as 16 times around the Earth (the number of circuits varying with flight altitude), then in the course of one single Earth day, the satellite can have just as many day-night cycles as it has trips around the Earth.

Night on the satellite is a sort of solar eclipse. The Earth's shadow covers the Sun at night on a satellite. Night descends when the satellite enters the Earth's shadow, but since the shadow covers only a small part of its orbit, the satellite night is shorter than its day (Fig. 20). Thus, for instance, on an Earth satellite doing 16 trips around the Earth per sidereal day, the satellite day lasts one hour, 29 minutes and 45 seconds, and its longest "winter" night is 37 minutes.

Twilight will precede the night on an artificial satellite just as it does on Earth (Fig. 20). There will be twilight on a satellite also before day breaks. But the satellite's evening dusk and the half darkness before dawn will be caused by quite different factors than those prevailing on Earth, where the sunlight is dispersed by the upper layers of the atmosphere. Satellite twilight will result from its passage through the penumbra of our planet. The satellite first enters the Earth's penumbra and only after that plunges into the darkness of its full shadow cone. The Earth's full shadow cone (umbra) gradually declines in diameter with distance into space, and finally runs out to an end, but its penumbra is continuously expanding outward in diameter. During a satellite night, the Sun will disappear, but it will be partially visible in twilight on the satellite.

The wonderful view of the rising and setting Sun always delights us. The extraordinary colors of sunrise and sunset are caused by the solar rays passing through thick layers of air. The colorful effects of the Earth's atmosphere will be still more intense during sunrise and sunset as they are beheld on an Earth satellite, because before

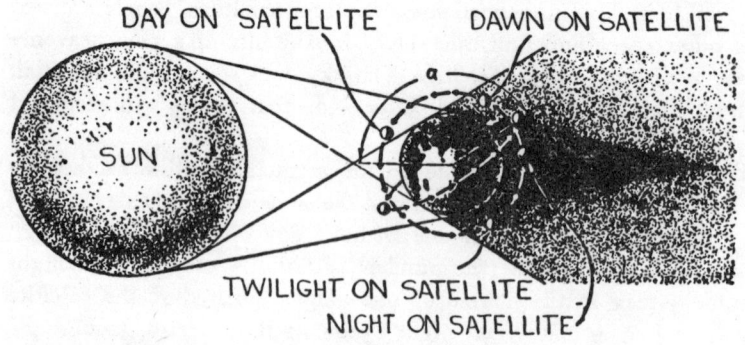

Fig. 20 Dawn, day, evening twilight, and night on Earth satellite. a is the inclination angle of the satellite's orbital plane with reference to the direction of solar rays.

reaching the observer's eye there, the solar rays will pass twice as far through the atmosphere.

An artificial satellite will also have its own seasons of the year, which, like those on Earth, will be conditioned by changes in the duration of night and day. The variations in the length of day and night, however, will not be caused by the same factors that prevail on our planet. On Earth it is the inclination of the Earth's axis to the ecliptic that causes the variation in the length of day and night during the course of the year; but on an Earth satellite these changes in daylight time depend on the varied length of time the satellite takes to pass through the Earth's shadow. On the satellite conditional winter comes at the period of the longest night, and summer in the season of the longest days.

The annual course of the satellite calendar is determined by its circular path, always lying in one and the same plane, which is fixed with reference to the stars. Let us take, for example, an Earth satellite in orbit over the poles at an altitude of 210 kilometers. The orbital plane of such a satellite will be, as the Earth's axis is, inclined to the ecliptic at an angle of 66° 33'. We assume that during the autumnal equinox this plane is parallel to the solar rays. At the moment the satellite enters the Earth's shadow, night falls on it. This happens four minutes after the satellite passes over the North Pole, and during these four minutes the satellite will still be flooded with Sun rays, although night already envelops the polar regions underneath it. But when the Sun rises on the satellite, the Earth's Antarctic surface will still be in the shade, and this will again last four minutes, until the satellite reaches the South Pole. Thus, day on the satellite will last 16 minutes longer than night, its total length amounting to 52 minutes. The satellite will at that time have its longest night and, therefore, its winter season.

On the above described satellite summer will be in full swing in June and December, and midwinter will be the end of March and September. During one year on Earth, the satellite will thus have two winters and two summers.

Knowledge about the respective length of night and day and annual seasons on an artificial satellite has an important bearing on observing it from Earth and for study of solar radiation by means of satellite-borne instruments.

CHAPTER FOUR
Utilization of Artificial Satellites

Space Ship Observatory and Laboratory

Unlimited expanses of the globe can be explored over a long period of time by means of artificial satellites, in contrast to the explorations possible with high altitude rockets, which are limited in space and restricted to a few moments of time. The artificial satellite thus combines the qualities both of baloons capable of remaining above the Earth for a long time and of rockets capable of ascending very high.

Artificial satellites will be useful first of all in the capacity of space ship observatories for study of the Earth's surface. On a satellite space ship can be installed precision instruments which will automatically make observations of natural phenomena taking place in the upper layers of the atmosphere and in outer space. These unmanned vehicles will record the results of their measurements and transmit the data by radio to Earth. Our knowledge about cosmic space will thus be enriched by much data that cannot now be gained by instruments flown in high altitude rockets.

The second Soviet satellite is precisely such a space ship laboratory.

Exploration of the Globe

Objects on Earth four meters in diameter will be visible from a satellite space ship orbiting at 200 kilometers altitude, when viewed through a 15-power prism binocular. Owing to the satellite's swift travel, however, it will be necessary to fixate the image with the aid of a special mechanism. To distinguish details of the Earth's surface at the horizon will be extremely difficult, however.

From a satellite space ship man will for the first time be able to

Fig. 21 View of the Earth's surface from an altitude of 225 kilometers

see the globe hanging in space. But what the Earth looks like from such a satellite's flight altitude we already know from rocket photographs like that shown in Fig. 21. This shows the appearance of a section of the Earth's surface as seen from an altitude of 225 kilometers. Since the photographing was done through almost the whose mass of the atmosphere, an infrared filter was used. On the photograph are reflected various details of the Earth's surface, cumulations of clouds, layers of atmosphere at the horizon and the curvature of the Earth.

An observer will see approximately the same picture from an orbital space ship.

Accurate mapping of continents has now been completed for only seven per cent of the Earth's land surface. From an earth satellite, meanwhile, it will be possible without special effort to carry out aerial photographic mapping of places difficult of access to surveyors at present, and to make more precise maps that have become obsolete because of changes brought about by the construction of airports, roads, dams and so forth. It is true that the safe automatic return of exposed film to earth will be required, if such surveying is to succeed. How this problem can be solved is as yet unclear.

The number of aerial photographs needed to map the entire globe grows less with increase in the satellite's flight altitude. The entire surface of the Earth can be photographed in daylight in less than 12 hours from a satellite in orbit at an altitude varying within the limits of a few thousand kilometers.

Great expanses of Earth will certainly always be covered with clouds, but in this case photographing can be done in the invisible infrared rays and also by means of radar equipment. Even when clouds are entirely absent, in a perfectly clear sky, the air still remains somewhat hazy, anyway. Photographs taken from an Earth satellite high above the atmosphere will nevertheless be extremely accurate. The distortions caused by the atmosphere when looking through a telescope, will be practically imperceptible, just as the text of a printed sheet is easily read through wax paper closely pressed over it, whereas letter outlines completely dissolve if the wax paper is held close to the eyes. The presence of cosmic rays and also X-rays in outer space is, we note, a serious complication in the technology of photographing from an artificial satellite, because ordinary adapters afford no guarantee against premature spoilage of photographic film.

Television transmitters on artificial satellites of the Earth or Moon can transmit to survey offices ordinary or stereoscopic pictures of the Earth or Moon surface visible from the satellite orbital altitude. In the making of stereoscopic pictures, because satellites are small in size, the two transmitting cameras will have to be installed on two satellites travelling approximately equidistant from each other. (Stereoscopic photographs of the Earth's surface can also be obtained from one satellite, by having contrasting photographs taken with one camera at varied times.)

An artificial satellite will also make it possible to measure the radiation and albedo of the earth (albedo is measurement indicating how much sunlight the planet or satellite reflects). The Earth's albedo varies within very wide limits chiefly because of irregularity in its cloud cover. From an artificial satellite this magnitude can be easily determined for various latitudes and seasons of the year.

Accurate observations of the orbits of Earth satellites even those minimal in size will, some specialists think, afford an opportunity to make various measurements of the earth as, for example, triangulation of the globe (method of measuring the earth's surface by plotting a set of triangles), especially of water bodies, measurement of distances between continents, etc. With the help of an artificial satellite, the width of the Atlantic Ocean can be measured to an accuracy reaching 30 meters. By means of an artificial satellite, the hypothesis about the relative migration of continents can thus be finally proved or disproved.

Surface depression in the oceans, i.e. lowering of a water body's real sea level as compared to its theoretical level, is known to reach several hundreds of meters. Grounds also exist for assuming that the shape of the Earth is itself gradually changing. All these changes can be kept under observation from observatory space ships.

Polar aircraft maintain a constant watch of ice-floe drifts in arctic seas, and forest rangers in fire patrol planes fly over the vast timberlands. Such lookout duty could, however, be done a great deal more effectively from artificial satellites. Satellite instruments can forewarn navigators at sea of floe ice jams. All icebergs, particularly of dangerous size, will be kept on record and informed sea captains can avoid shipwreck on drifting icebergs in the high seas. The spaceship observatories can report to Earth on the rise of forest fires in the depth of the taiga wilderness and indicate exactly where the fire is located.

Earth satellites may possibly be suitable for coast guard rescue service, sweeping mine fields in naval ranges, etc. Perhaps a means will be found to use them in forecasting the runs of fish shoals. Likewise, from an artificial satellite, it may prove possible to fix the location of sunken ships and wrecked airplanes, and perhaps also the sites of lost expeditions.

Opinions are advanced that oceanographic, glaciological (glacier) and seismological (concerning oscillations of the Earth's crust) investigations can in future be made by means of an artificial satellite.

Glaciologists are hoping, for instance, that observations from the artificial satellites will confirm their hypothesis about a gradual change in the Earth's climate, which is thought to be moderating and to involve the gradual thawing of our planet's polar ice caps.

Exploration of the Atmosphere

Artificial satellites will be useful in meteorological observations. They will permit tracking the distribution and shifting of cloud formations, to determine the character of the Earth's cloud cover, the boundaries of warm and cold air masses and the spread of storms. Weather observatories are currently too few in number even on land which occupies less than 30 per cent of the Earth's surface. The tens of thousands of meteorological stations that exist are found incapable of creating, for instance, a complete picture of the Earth's cloud cover even over the continents, not to mention the immense water expanses of the globe. Specialists are of the opinion that from the altitude of an artificial satellite it will still be difficult to determine the movements of large cloud formations inasmuch as continental boundaries, which must necessarily be known for this purpose, will themselves be obscured by the clouds. It will therefore be difficult to determine how much massive cloud formations have shifted and to what degree their dimensions changed during one satellite trip around the Earth.

The density of the atmosphere's upper layers can also be determined indirectly through Earth satellites which aren't even equipped with any instruments. Sufficient for this purpose will be visual or radar observations of such a satellite's orbit and the degree to which its travel is braked by the atmospheric friction.

The satellite can periodically eject sodium vapors which will shine brightly in the solar rays. From the process of the sodium trail's spread, the temperature in the upper layers of atmosphere can be judged. The curving deformation of the sodium "cloud" will also serve as a way to ascertain wind velocity at the given altitude.

Just as remote-control meteorological stations can, with appropriate instruments, measure weather elements at a distance, so the temperature, pressure and air density at various altitudes can be determined by means of artificial satellites. A polar satellite in particular will afford the possibility of rapidly finding parameters that characterize atmospheric conditions and other data at a constant high altitude along the meridians.

From the first Soviet satellite, its own temperature and other data were transmitted to earth through a certain variation in the duration of signals and pauses between them, which on the average lasted 0.3 seconds.

All the foregoing shows what great importance artificial satellites will have for accurate weather forecasting.

It has been established that the Earth's atmosphere gleams at high altitudes. In daytime even at an altitude of 120 kilometers the sky is not black. At such an altitude the sky's luminosity amounts to four per cent of what it is in the zenith at sea level. The sky continues to be luminous even at night.

Variation in the natural radiation of the Earth's atmosphere as it depends on season of the year and geographic co-ordinate can be investigated from on board a satellite in orbit around our planet.

Exploration of the Ionosphere and Radio Wave Propagation

By means of artificial satellites, the atmosphere's ionization (propagation of ions and electrons) can be investigated at various altitudes. Ionization studies will contribute very much, particularly, to forecasting radio communication conditions.

Since the distance between artificial satellites and the radio receiving station on Earth will be changing constantly and the thickness of the air layer dividing them will be now increased, again decreased, the quality of ions situated between the radio transmitter and the radio receiver will also be variable. Accordingly, the character of radio signals received from the artificial satellite will also vary during its various positions with reference to the receiving station. Analysis of these signal variations will permit judging what communication conditions are like in the ionosphere.

Comparatively short range radio stations on board ships at sea will be able to maintain ship-to-shore contact with the mainland through a satellite space ship which can daily appear above the horizon, extending the communication range by relay. Such communication can be accomplished, we note, also by light signals, which penetrate the atmosphere a great deal easier when they are sent upward, since their travel distance in the absorbing medium is thus shortened.

The artificial satellites can also serve to relay the ultra-short waves of television broadcasts in particular, thus extending the TV broadcast range over long distances. Because of equipment complexity and the need for powerful sources of electric energy, such an application to TV is certainly not immediately envisaged, but it is not ruled out that in future satellite space ships will be found to be a profitable means for the transmission of television broadcasts from one continent to another.

Exploration of the Earth's Magnetic Field

The magnetic field of our planet has been fairly well explored at the Earth's surface and the results of these investigations have long since been utilized in sea navigation, aviation, geodesy and other fields. The magnetic field of the Earth is made up of a permanent field, created by sources within the Earth, and a variable field created by electrical currents circulating in the ionosphere and space beyond the atmosphere. The periodic regular oscillations of the Earth's magnetic field are daily, 27-day (connected with the period of the Sun's rotation around its own axis), annual, 11-year (connected with the cycle of solar activity) and secular. Sudden variations of the Earth's magnetic field (magnetic storms) are also observed. It is conjectured that the terrestrial magnetic field has a deflecting effect on charged particles travelling around the Earth. On the other hand, it is assumed that oscillations in the Earth's magnetic field rise in consequence of charged particles from the Sun penetrating the Earth's atmosphere.

By means of artificial satellites, especially satellites circling in elliptical orbits, a magnetic survey can be made of the space surrounding the Earth, to study the causes of the magnetic anomalies—deviations in the intensity of the Earth's magnetic field from the average ("normal") magnitudes for a given locality. It will be possible to investigate the effect that electric currents originating at very high altitudes have on the Earth's magnetic field, and to study the effect changes in the intensity of cosmic rays have on the course of magnetic storms. Exploration of the Earth's magnetic field by means of artificial satellites will have not only a theoretical, but also a practical, value. Such investigations will, for example, open the way to discovery of useful mineral deposits and to assessment of their reserves.

Biological Investigations

The artificial satellite is also of interest from the viewpoint of studying the prospects of interplanetary flights. On a satellite space ship can be studied what effect weightlessness or zero gravity has on physiological and psychic processes, and also how cosmic, solar and other radiations affect living organisms not protected by the Earth's atmosphere. Such experiments were made on the second Soviet Earth satellite. On a satellite can be verified the hypothesis advanced by K. E. Tsiolkovsky that in the absence of gravity, plants and organisms from the simplest to the most complex will grow and develop a great deal faster than they do where gravity exists.

Meteorites, Micrometeorites and "Cosmic Dust"

Micrometeorites, it is assumed, affect in some degree the conditions of the ionosphere and thereby the propagation of radio beams.

Devices to count micrometeorites can be installed on satellites; the distribution of these particles, their impetus and electrical charge depending on geographic latitude, can also be determined. The crackle of micrometeorites hitting the satellite plating can be picked up by a crystal microphone on board the space ship and transmitted to the earth after preamplification of the radio signal.

The well polished plating of the satellite space ship will gradually get tarnished from being hit by micrometeorites. And this effect can prove to be a substantial help in the study of micrometeorite characteristics. The satellite plating can also be covered with a radioactive coating, and through telemetering of the satellite's radioactivity, the extent to which meteoric dust and micrometeorites rub off this coating can be judged by the gradual decline in the intensity of its radioactivity.

The satellite space ship can be made completely airtight and before launching be filled with a gas under pressure. This will give it the rigidity needed to withstand the high overloads that develop during launching. Then, when coasting on its own momentum, a drop in gas pressure will indirectly point to a puncture made in its plating by a meteorite. The length of flight without collision will give an indication on the possible frequency of meteorite hits, and the rate in gas pressure drop will obliquely point to the meteorite's size and travelling speed.

From an artificial satellite, meteors will be visible not on a sky background, but against the earth plunged in the darkness of night. These new conditions for observing meteors may, perhaps, enlarge possibilities for studying them. It will become possible on board a satellite to collect specimens of meteoric dust and ascertain what influence it has on the formation of weather.

"Cosmic dust" is also to be observed in outer space. It is sometimes successfully discovered at the very surface of the Earth as well. Artificial satellites can also be used to study cosmic dust. These particles of interplanetary dust will present no obstacles to the travel of a satellite space ship.

Artificial Meteors

Artificial meteors of definite form and composition can be thrown off a satellite. This meteor-throwing can provide a wealth of material for the study both of natural meteors and of conditions for braking space ships by atmospheric means. If from a satellite in orbit at an altitude from 200 to 1000 kilometers, a meteoric body be launched with a speed of 50 to 250 meters per second (in a direction opposite to satellite travel), this will suffice to make the artificial meteor cut into the atmosphere with a velocity of about eight kilometers per second. And, what is very important, in each separate case in which a meteoric body is thrown from a satellite not only will the speed at which it invades the atmosphere be known, but also the path over which it travels from the moment of launching to its penetration into the atmosphere. All these data and the meteor launching time can be communicated in advance to observatories on earth. Attempts have already been made to make a photographic record of the path travelled by artificial meteors which were developed by throwing metal balls down from high altitude rockets.

Atmospheric Observations

The aurora polaris and also the zodiacal light (observable in the form of a feebly shining cone on the night sky background at a certain season of the year before sunrise or sunset in the area of the zodiacal constellations, that is, along the ecliptic) are more readily investigated in outer space beyond the Earth's atmosphere, because on Earth the luminosity of the upper atmospheric layers distorts the normal picture of these natural phenomena. Detailed exploration will become possible also of what is called the "gas tail of the Earth"—a long projection of the top atmospheric layers on the side of the Earth approximately opposite to the Sun.

Even at night the atmosphere hinders the photographing of very feeble celestial objects by means of astrographs, while in daytime the Earth's air envelope makes observation of the starry sky wholly impossible. But the atmosphere no longer causes distortions at the flight altitude of an artificial satellite. This creates conditions especially favorable for astronomical observations. Stars that do not twinkle are a great deal easier to observe and photograph. Photographs of the planets and their satellites can, in such conditions, be taken with any magnification, whereas at observatories on earth a magnification of one thousand times is already difficult, owing to the optical "whirls" caused by the atmosphere. On an artificial satellite, moreover, astronomical optical observations will not depend on whims of the weather.

Possibilities for radioastronomy are also considerably enhanced on space ship observatories, since many radio beams from outer space which never reach the Earth's surface can be detected before they get to the atmosphere.

In the future artificial satellites will have electronic equipment with television transmitters, thanks to which observers on Earth can "look" at the sky through a space ship telescope.

An artificial satellite can also serve in outer space study of cosmic rays, for instance, to determine their content of nuclei of lithium, beryllium, boron and other elements. Of great importance also are investigations of the variation in the intensity of cosmic radiation (and the ionization created by this radiation) depending on time, altitude and geographic co-ordinates.

Selection of Essential Orbits

Satellite space ships can, as we see, find the most varied application. The study of various phenomena will evidently require, however, that satellites travel in specially selected orbits. It is clear, for instance, that equatorial satellites are as unsuitable for investigating aurora polaris as polar orbits are useless for the study of the zodiacal light.

Artificial satellites at variable altitude will have special application. Travelling in elliptical orbits, such satellites will oscillate in altitude, now ascending to more rarefied layers of the ionosphere, again descending to denser layers of air. This will permit making observations at various altitudes from one space ship and will fill out gaps that exist in our knowledge of solar radiation, the atmosphere's composition at various altitudes, the ozone distribution, the Earth's magnetic field and ionospheric storms.

The southern hemisphere has little land and much water space. It will be of interest, therefore, to select a satellite's orbit so that the space ship observatory will spend more time over the northern hemisphere and less over the southern. This can be done by lengthening the orbit's northern section, which will in addition reduce apparent travelling speed in this orbital section. In other words, the satellite should have an elliptical orbit with its perigee over the southern hemisphere. Besides, the higher the orbit's apogee is, the less time will the space ship spend over the southern hemisphere and the more time over the northern.

Prospects

Many of the problems enumerated above are only in elementary stages of study and still require many years of investigations.

The use of satellite space ships will no doubt lead to the future discovery of natural phenomena, existence of which we do not at present even suspect.

Many years will doubtless be spent in exploring all the above-listed problems. In the course of the first years, therefore, artificial satellites will be applied to study a narrower circle of problems.

Satellite as Interplanetary Space Station

From the viewpoint of astronautics, the use of artificial satellites in the capacity of interplanetary space stations has the greatest importance.

To reach the Moon, Venus, Mars—our nearest neighbors in outer space—the interplanetary space ship must develop at take-off a speed over thirty times greater than the speed of sound. It is beyond the possibilities of contemporary technology to construct such an interplanetary ship. For easier solution of this problem, the cosmic voyage can be divided into stages, as K. E. Tsiolkovsky proposed in the Eighteen-Nineties. For this purpose an artificial Earth satellite can be employed as a kind of transfer station in space.

At transfer points in transportation systems on Earth during stops at stations, ports and airdromes, ships and locomotives fill up with supplies of coal and water, airplanes refuel with gasoline, passengers stock up with food. Sometimes a new locomotive is coupled to the railway train, one airplane replaces another. The building of an interplanetary transfer station will have similar value for a flight into outer space. Such a station might serve as a springboard for man's further penetration into the universe. At the transfer station space travellers can stock up with everything necessary for the continuation and completion of a long cosmic voyage: fuel which the space ship could not carry along when starting from the Earth's surface, equipment, provisions and so forth.

The space ship and also the diversified payload it needs to reach the journey's terminal destination, can be delivered in advance separately to such a space station. This will simplify space ship design, since in taking off from a satellite platform a much smaller supply of fuel will be needed than would be required were the take-off directly from Earth.

When people travel on Earth and stop at an intermediate station, they come to a complete standstill. It will not be so with space travellers who stop over at an interplanetary station. They will acquire both the orbit of the station and its orbital velocity and, on leaving, take the speed along.

The space station can, for cosmic ships of certain types, also be useful for the return to Earth, as a transfer point where the Earth-bound crew board the space glider on which the descent to land on our planet is made.

Satellites which fly over the poles and are convenient for observation purposes are not as a rule suitable as interplanetary stations, and this is why: The interplanetary station should necessarily travel together with the Earth in the plane in which our planet rotates around the Sun (the so-called plane of the ecliptic; all other planet's of our solar system also travel roughly in this same plane). Only in this case will the space ship setting forth from the interplanetary station have a travel direction more or less parallel to the direction of the Earth's travel in its own orbit around the Sun. And this is extremely important in taking off from the transfer station on a flight into outer space, because the Earth's orbital speed will then be added to the ship's own take-off speed, assisting it to overcome the pull of gravity from the Earth and the Sun.

The orbital mobility of the intermediate space station is its principal advantage for realization of space flights. Thanks to the station's swift motion, the rocket on landing on it retains its speed and makes use of this speed in the take-off for further travel in outer space. Thus, estimates show, for example, that in taking off from an artificial Earth satellite on a trip to the Moon, Venus or Mars, the interplanetary rocket will have to develop a speed of but 3.1 to 3.6 kilometers per second instead of the 11.7 to 11.6 kilometers per second that would be required if the take-off were directly from the surface of the Earth. This advantage comes from the transfer station itself having a speed of about 8 kilometers a second. This means that a rocket capable of ascending from the surface of the Earth to an altitude of 1000 kilometers (and such flights have already been made) would already be able to reach Venus or Mars, if it started from an interplanetary space station.

In most space travel projects it is foreseen that at the interplanetary station astronauts will transfer to a ship which has been assembled in station workshops from parts delivered from the Earth. Engines and also other units taken from rockets that have arrived at the station can be utilized in equipping the interplanetary ship. The flight conditions on the trip from Earth to the artificial space station are quite different from what they are on the trip from the satellite station into outer space. Tne rockets for these interplanetary voyages must also, therefore, have different designs.

For the flight from the Earth to the satellite space station, the space ship must be streamlined, since it has to fly through the entire mass of the atmosphere. It must be equipped with a powerful engine capable of communicating to it a speed of about 8 kilometers a second, and consequently also a relatively large supply of fuel for feeding the engine. But for the flight from the satellite station into interplanetary space the ship can have a shape that is not streamlined, since in outer space it will not encounter resistance of the material environment. Owing to this, the fuel tanks can be made spherical, which will reduce their weight for a given volume.

The start from a satellite transfer station can be made with rockets of considerably less power than is needed in the take-off from the surface of the Earth. In fact, when starting from the Earth's surface the engine pull must be greater than the rocket weight, whereas there is no such requirement when taking off from a satellite space station. Even with engine pull somewhat less than the rocket weight on Earth, the ship can gradually gain the necessary speed. In the take-off from Earth the greater part of engine power is spent not only on useful work in overcoming terrestrial magnetism, but also on various losses (air-drag, for example), and with cessation of engine work the rocket usually falls back to the Earth. But such a danger does not threaten the ship flying from a space station. Even if the rocket engine ceases to run, the ship does not fall either back to the departure station in space or to the Earth. The ship will therefore have to take along an incomparably smaller quantity of fuel than it would in a start from Earth. This is still another reason why it is advantageous to use an artificial satellite in the capacity of interplanetary station.

According to some space travel plans, the rocket which arrives from Earth at the interplanetary station will itself serve as the vehicle for the further flight into outer space, after its streamlined plating has been removed at the space station. Its air fins with rudders will no longer be needed. If flight direction has to be changed in outer space, a gas jet will be emitted from the rocket in the direction wanted. After refilling with fuel at the transfer station, the rocket sets forth on further flight in space. The more fuel the rocket takes along on leaving the space station, the more outlet velocity will it develop, of course. This is, however, not always the case when taking off from Earth. In taking off from a space station extra fuel always yields a positive effect; but in starting from the Earth itself extra fuel can even have a negative effect (less speed and altitude) because of excessive rocket load.

The satellite space station is, however, not always a necessary stage enroute when making a flight to the Moon and planets. Such a flight can also be undertaken without stopping at an interplanetary station. In this case the take-off will be somewhat different. The rocket will start from the Earth's surface and after developing a speed of about 8 kilometers a second, be converted into an artificial satellite at an altitude of 200 to 300 kilometers. Auxiliary rockets will gradually deliver to such a rocket-satellite the additional payload and fuel necessary for further travel in interplanetary space. After receiving "reinforcement," the interplanetary rocket will set forth on a trip to its planned destination in space. Such a solution, involving use of a temporary artificial satellite, is of interest as a way of curtailing the meteor hazard that could make things difficult on a permanent space station.

For a long time still before man flies forth into the limitless expanses of the universe, the conditions of such flights can be tested at an interplanetary station. On it can be determined whether or not prolonged absence of gravity is harmless for the human organism, how artificial gravity affects man and so forth. On such a celestial island protective means against the meteoric danger can also be successfully studied. Using the interplanetary station as a base near our planet, astronauts will be able to learn the intricate art of ship navigation in airless space, and also to master the skill of braking cosmic speed in gliding flight on descent to the Earth.

At a space station also many facts can be determined that are essential for creating the most rational design of space ship and glider.

The use of an artificial satellite in the capacity of interplanetary transfer station or, what is the same in principle, the conversion of a space ship temporarily into an Earth satellite, will probably be practised only during the elementary stage in the development of the technology of cosmic flights. The powerful atomic ships of the future will not, enroute to the Moon and the planets, have to shift into a circular orbit and get "reinforcement" from Earth. Possibly also the dispatch of small multi-stage guided rockets to the Moon and the planets will prove simpler to accomplish with take-off directly from the Earth's surface.

Natural Interplanetary Stations (Moons)

Assertions that the Moon can be used as an interplanetary transfer station are encountered in astronautical literature. The Moon is, however, unsuitable for this purpose. It is situated too far from the Earth's surface. Moreover, inasmuch as the Moon's mass, and consequently also its attraction, is comparatively great, much fuel would have to be spent first, for braking during the space ship's descent onto lunar surface and, after that, for the take-off from the Moon.

Let an expedition set forth, for example, to Mars. Calculation shows that if an an artificial satellite situated a short distance from the Earth were used in the capacity of transfer station, then on the whole trip during the flight from the Earth to the station and from the station to Mars the space ship will have to develop less total speed (and consequently, consume less fuel) than it would just for the flight to the Moon alone. The explanation of this is that in the descent to the surface of Mars the braking of the ship can be done with the aid of the resistance of the planet's gas envelope, whereas since an atmosphere is lacking on the Moon, it would be necessary to use for this purpose the energy of the rocket engine.

The use of the Moon as an interplanetary station would make some sense only in case especially high-grade fuel and construction materials were to be found on the Moon.

So an artificial Earth satellite has a number of advantages over the Moon in the matter of serving as an intermediate interplanetary station. In the first place, the satellite station can be located sufficiently close to the Earth, which will make it possible to fly there a great deal faster and with less fuel consumption. Secondly, since the satellite lacks its own field of gravitation, fuel can be saved which would have to be consumed for making a landing on the Moon and subsequently taking off from its surface.

But does not the Earth have a second moon or even a few such natural satellites that are located closer to the earth than is the Moon, the only satellite we know—satellites which have heretofore been unobserved? Some other planets of the solar system have several moons. For example, Jupiter has 12 and Saturn has nine satellites. The dimensions of many satellites of other planets are extremely small. The diameters of the Mars satellites—Phobos and Deimos—are respectively 14 and eight kilometers. A second natural satellite of the Earth, even if it were extremely small, would be an important way station base for man's invasion of space. The discovery of such a satellite (or several such satellites) would make it considerably easier to solve the problem of flight to the Moon and the planets, rendering superfluous the building of an artificial satellite.

Both a flying observatory and an interplanetary station are comparatively easy to outfit on a natural satellite.

It stands to reason that even if there are such satellites of the Earth, they can be only quite tiny and discovering them is exceedingly complicated. Owing to its high travelling speed such a tiny satellite cannot be detected in a telescope, and is even less visible if orbiting close to the Earth. Moreover, if sufficiently near the Earth its passage can leave no traces on photographic plate because the exposure time would be too short. In addition, on getting into the Earth's shadow, such a satellite does not shine, and observation of it will therfore be confined to a brief interval of time. Astronomers admit the possibility that such a satellite could have at some time been observed but mistaken for a meteor. The recently developed methods of radio astronomy applied to investigate meteors may be useful in solving this problem. Observations in this direction are being conducted, for instance, by the Meteorite Institute of New Mexico (USA) under the management of C. W. Tombaugh, who in 1930 discovered the planet Pluto.

It is evident that if new moons are discovered they will be outside the limits of the atmosphere. Otherwise they would have long since fallen to the surface of the Earth or have burned up from air friction.

The natural interplanetary stations of other planets of the solar system are also of great interest for astronauts. Thus, a journey to Mars, for example, with descent to its surface will evidently be preceded by reconnaissance flights around that planet. (A flight to the Moon is apparently a similar matter.) For this purpose the rocket ships are converted for a time into artificial satellites of Mars. As a matter of fact, a landing on the planet Mars with subsequent take-off will involve enormous difficulties in the early eras of space travel, the more so since all the fuel necessary for the return voyage will have to be taken along on the outward journey from earth.

CHAPTER FIVE

On the Space Ship

Take-off

An automobile, a train and a sailing vessel continue to move as long as they are driven by an engine or the wind. But as soon as the engine is stopped or the sails furled, any form of motion ceases.

It is true that they do not come to a standstill at once but continue moving under their own momentum for some time. However, they cannot get far in this way, since the amount of energy they have accumulated will soon be neutralized by friction and air resistance.

The situation is quite different where the space ship is concerned. In a few minutes its motors will give it great velocity and the rocket will cover the remaining part of its journey under its own momentum, since it will meet neither friction nor air resistance in space.

The sooner the space ship attains the required speed, the less time will it need to overcome the force of gravity and the less fuel will it burn.

A great amount of fuel would be saved if the rocket could reach the required speed instantly and then continue on its route under its own momentum. However, that is a practical impossibility—the rocket can only gain speed gradually as the fuel burns. Besides, the initial speed must not be greater than the human organism can endure.

Very often on the covers of books dealing with interplanetary travel one can see a space ship flying along a straight line between the Earth and the Moon. It has covered half the distance or is even approaching its destination with its motors still in operation. This conception is quite incorrect. The trajectory of the space ship can never be a straight line, and its motors must be shut down a few minutes after the take-off, at a short distance from the Earth. Only in this way can the space ship save enough fuel for the return journey.

A successful flight will depend in large measure on whether the correct trajectory is chosen. Trajectories involving the minimum expenditure of fuel are very intricate—the rocket must constantly change direction and acceleration. If a simplified trajectory (for example, a vertical one) is chosen, the fuel consumption will be several times as great.

The timing of the take-off is of paramount importance for the whole undertaking, since neither the Earth nor the celestial body for which the space ship is bound is stationary in space.

Something similar happens in shooting from guns mounted on a battleship rocking in sea waves. The weapon is loaded. The gunner is at the range-finder sights. For a moment the target flashes in view. The gunner swifty fires; if an instant late, the shell digs into the waves at the ship's side or flies high into the sky. The take-off of a space craft will consequently differ radically from an aircraft take-off.

Flight

Once the motors are cut out the space ship will coast the remaining distance between the two planets (more than 99 per cent of the whole distance). For example, rockets setting out for neighbouring celestial bodies will use their motors for only the first 2,000 kilometers or so of their journey, whereas the distance to the Moon is estimated in terms of hundreds of thousands of kilometers and to the planets in terms of millions of kilometers.

On the Earth only railway transport moves along a strictly defined line; other types of transport continually deviate from the geometrical line of their route because of the imperfections of roads, the influence of air and water currents, uneven operation of their motors and other reasons.

In space things are quite different. Along its entire route the space ship will be influenced only by the attraction of the Sun, and it will speed along a strictly defined line as if it were on an invisible railway line.

It might seem that a slight deflection from its trajectory would not be very dangerous for the space ship, since it would have enough room in space to avoid collision with other space ships. However, piloting in space must be carried out with greater accuracy and vigilance than at sea or in the air. The slightest deviation in speed or direction may result in serious consequences, as can be seen from the following examples.

A space ship flying to the Moon with a minimum initial speed would fall four thousand kilometers short of its destination, were its speed reduced by just one meter per second. One can judge how difficult it would be to pilot a space ship if its acceleration in space were four to five meters per tenth of a second.

It is an even more serious matter where flights to the planets are concerned. A reduction of one meter per second in speed would be followed by a reduction of tens and perhaps hundreds of thousands of kilometers in the rocket's range.

Let us assume that we are leaving for Jupiter on a trajectory which requires the lowest take-off speed of 14,226 meters per second. If this speed is reduced by one meter per second, the space ship will fall 400,000 kilometers short of its destination. If the error is 0.1 per cent the ship's deflection from its target will be measured in terms of more than five million kilometers.

This occurs because at a great distance from the Earth and the Sun the force of gravitation is quite imperceptible and the least increase of speed substantially lengthens the rocket's flight range.

A deflection of 0.1° from the departure angle may result in the rocket missing the target by one million kilometers.

To avoid such errors pilots will have to take constant observations and to adjust the rocket's course by switching on and off a low-powered steering motor.

How are astronauts to measure the distance they have covered?

When flying to the Moon this will be done by ascertaining the angle at which the Earth or the Moon is visible: the lesser the angle, the greater the distance. The distance from the Sun will be reckoned by changes in temperature—modern electrical thermometers can detect temperature vacillations of $0.000001°C$. Such instruments will make it possible to compute the distance from the Sun within two to three kilometers.

Descent

How will the descent of the returning space ship be effected?

Theoretically, a rocket motor could be used for the purpose. Turned round to point towards the Earth, it would decrease the rocket's speed through the action of the exhaust gases pushing the rocket in the opposite direction. But the amount of fuel required for the operation would be enormous, and no rocket would be large enough to take it.

The utilization of air resistance is another method of slowing down the space ship. However, the inevitable frictional heating of the space vehicle when flying through the atmosphere at cosmic speed cannot but cause misgivings. The example of meteors, "shooting stars", which get white hot on penetration into the atmosphere shows that a space vehicle's descent from cosmic space to the Earth is a complex problem. Frictional heating will make it impossible to use parachutes which would burn instantly. The same is true of the space ship flying from an artificial satellite. It will be altogether unsuitable for landing on the Earth, being a bulky structure with thin walls without any streamlining. Once in the atmosphere it will soon become white-hot. On approaching the upper strata of the atmosphere the crew will therefore have to take their seats in a perfectly streamlined space glider. As far as the space ship itself is concerned, it will either burn in the atmosphere like a meteor or become an artificial satellite of the Earth, if there is enough fuel to settle it in a circular orbit.

The space glider will enter the highest levels of the atmosphere at a speed exceeding eleven kilometers per second and will emerge again into space after experiencing a certain amount of retardation due to air resistance. In this way, after a series of such maneuvers, the space-glider could shed most of its excess speed and avoid being heated up to a dangerous point during descent.

As the speed of the glider decreases the surface of its "rudimentary" wings will become insufficient for gliding, and at this juncture the retractable wings will be brought into operation. When the glider's speed is completely neutralized it will land, after a descent of a few hours.

The procedure will be the same for astronauts returning to the Earth from a space station. In this case the glider will be "thrown off" the station by its low-powered rocket motor which will give a slight push in the direction opposite to that in which the station is moving. Having lost some of its former speed, the glider will gradually enter the atmosphere.

CHAPTER SIX

Space Voyages

A Trip to the Moon

The Moon will undoubtedly be the first objective in a series of space trips. Its distance from the Earth is about 384,000 kilometers, i.e., $\frac{1}{100}$ of that of Venus when the latter is at its nearest point to the Earth. It is a comparatively small distance even on the Earth. There are a lot of railwaymen and sailors who have covered the same distance, and many aviators have flown more than twice as far.

Man is capable of climbing the highest mountains. But would he be strong enough to get to the Moon if a ladder were to be set up between the Earth and the Moon?

Numerous experiments have shown that a full working day is required to climb to a height of 1,550 meters, and at this rate it would take 680 years to reach the Moon. But this calculation is only valid on the assumption that the whole trip is made under the same conditions and at the same speed as on the first day. The assumption is wrong, however. The longer the traveller went on climbing, the more the Earth's gravity would decrease. This would enable him to walk faster and faster and to complete his journey in eleven years.

Now what about a rocket? How long will it take it to reach the

Moon? If a rocket were to escape from the Earth at a speed of 11.1 kilometers per second, the minimal velocity, it would arrive at its destination in five days. But if it took off from earth with a speed of 11.2 kilometers a second, the ship would get to the Moon in 51 hours.

Not only the first satellites, but also the first Moon rockets will be radio-controlled. The radio messages they will send out will enable the scientists to follow their flight. The scientists will know the rocket has crashed into the Moon's surface the moment they see a flare-up of the flash-powder it will carry. The flare-up will be particularly clearly visible from the Earth if the rocket falls on the unlit part of the Moon's face. In addition to this, the rocket may scatter white powder over a considerable area, which will also be seen from the Earth.

At a later stage more powerful manned rockets taking off from an interplanetary station may become artificial satellites of the Moon and revolve around it for a long time without expending any fuel, and it will be very economical therefore to make a study of the Moon from a rocket of this kind.

Calculations show that a rocket weighing ten tons with an exhaust velocity of four kilometers per second will not need more than twelve tons of fuel to make a round-the-Moon flight, if it takes off from an artificial satellite of the Earth. If it were to take off from the Earth it would have to carry 150 tons of fuel. With an exhaust velocity of 2.5 kilometers per second these figures would be 25 tons and 840 tons. It is assumed here that the space ship attains full speed instantly, without expending additional fuel on overcoming air resistance.

As only one hemisphere of the Moon is visible from the Earth, investigation of the other hemisphere would be of great scientific interest. A flight over that hemisphere could be made at a time when its entire surface was lit by the rays of the Sun, i.e., at new moon.

It can be assumed that the hemisphere that is invisible from the Earth does not differ fundamentally from the other one and is likely to be just as dry, waterless and virtually devoid of any atmosphere. Travellers can expect to see large black places where there are valleys, so-called "seas," mountain ridges cut by deep crevices, mountains brightly lit at the summit and completely dark at the base, wide jagged circular protuberances, precipitous inside but sloping gradually towards their outer edge (the so-called "circuses"), chains of craters and dazzling strips of snow-white volcanic ash ("light rays").

Let us imagine that a space ship designed as in Fig. 11 (see page 48) has been launched from an interplanetary station to investigate the Moon (I).

During the flight of the space ship under its own momentum its speed will vary. Launched at a great velocity, the rocket will gradually lose speed like a stone thrown upwards. In five days the rocket will reach a point where it is affected by the Moon's gravitational field. As soon as this happens its speed will begin to increase, reaching 2.5 kilometers per second some scores of kilometers from the Moon's surface.

If the space ship is to become an artificial satellite of the Moon at a height of ten kilometers from its surface its speed has to be reduced to 1.7 kilometers per second, a circular speed for this altitude (Fig. 11, II). Its period of revolution will be one hour 50 minutes and its visible horizon 186 kilometers; objects on the Moon's surface measuring three meters or more will be visible to the naked eye.

The space ship will revolve around the Moon as long as required, without consuming a pint of fuel (Fig. 11, III).

When it has been decided to start on the homeward journey the astronauts will switch on the motors. Having increased its speed, the space ship will leave the circular orbit, while the detached fuel tanks continue on their old route (Fig. 11, IV). The automatic instruments installed in them will regularly communicate to the Earth the results of the various measurements by radio.

The space ship will descend in the way already described (Fig. 11, V). The space glider will land on the Earth with its wings completely moved out (Fig. 11, VI).

Reconnaissance flights around the Moon will be followed by flights on which a landing is made. Is it possible to make a landing on the Moon's surface without using fuel? Has the Moon an atmosphere?

Observations have shown that the Moon's atmosphere is very thin. According to preliminary data the mass of the air over each square centimeter of the Moon's surface is one two-thousandth of that of the Earth. The density of the atmosphere on the Moon's surface is the same as that of the Earth at a height of sixty kilometers.

In all probability it cannot be used to slow down the space ship's speed before landing, and rocket-braking will therefore have to be used for the purpose.

On the Moon, just as on other planets without an atmosphere, the space-travellers will have to stay in airtight compartments or put on space suits before stepping outside. In spite of this burdensome clothing the travellers will be able to move about easily because the Moon's gravity is only one-sixth of that of our planet.

To break away from the Moon's gravitational field one needs one-twentieth of the energy required for the same purpose on the Earth. Consequently, the speed of escape from the Moon will be considerably less than that from the Earth—to be precise, less than 2.5 kilometers per second, whereas modern liquid-fuel rockets are capable of greater speeds.

A Trip to Mars

A flight to Mars will also be of great interest. During the past three centuries this planet has attracted particular attention on the part of astronomers and other scientists because of its proximity to the Earth and similar natural conditions. The experts are no longer satisfied with studying the surface of Mars from images that appear tiny even through the largest telescopes.

A trip to Mars, like one to the Moon, with a landing on its surface, will probably be preceded by reconaissance flights around the planet. For this purpose space ships will temporarily become artificial satellites of Mars. Landing and take-off will, as a matter of fact, be extremely difficult in the early stages of space-travel, the chief trouble being that fuel for the return journey will have to be brought from the Earth. A detailed investigation of the surface of Mars will make it possible to select areas suitable for landing and to obtain data which cannot be established from the Earth but is necessary before the launching of an expedition that is to land on Mars.

The first thing that must be investigated is whether the structure and composition of the Martian atmosphere can be used to slow down the space ship. Such an investigation will also help to discover whether man can live on this planet and whether its atmosphere provides adequate protection against harmful radiations and the infinite number of "falling stars" bombarding it from outer space.

It has been discovered that because the atmosphere of Mars contains practically no ozone, which absorbs the ultra-violet rays of the Sun, these rays penetrate to the planet's surface and may endanger the lives of space-travellers.

Flights around Mars could be made along different trajectories, the duration of the trip and the initial speed of the space ship depending on the trajectory chosen.

Let us consider a trajectory involving a two-year journey (Fig. 22). The rocket will start from the interplanetary station at midnight

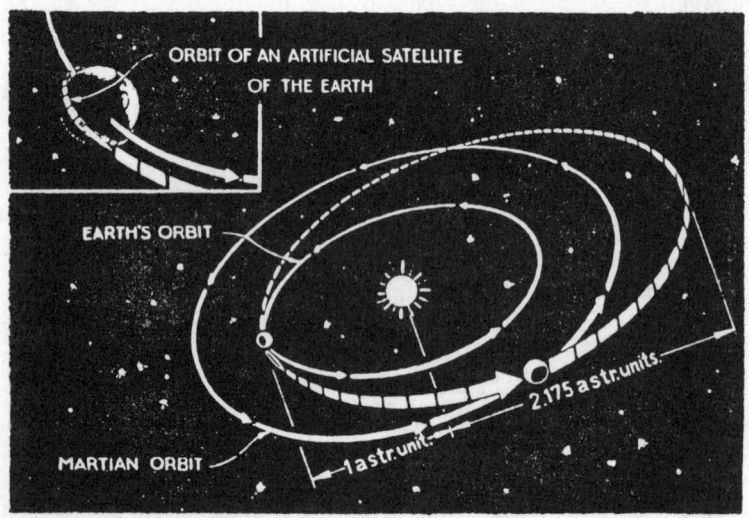

Fig. 22 Flight around Mars in two years. (Upper left) A rocket taking off from an artificial satellite of the Earth.

local time when the centers of the Earth, the Sun and the station are in a straight line. This is the most appropriate moment because the direction of the station's motion and that of the starting rocket will then coincide. Taking advantage of the speed of the space-station, the rocket will take off with the lowest speed of 4.3 kilometers per second, whereas a speed of 12.3 kilometers per second would be needed, were it to leave directly from the Earth.

A rocket weighing 10 tons with an exhaust velocity of four kilometers per second must have 19.6 tons of fuel on board if it is to take off from the interplanetary station, and 216 tons if it is to be launched from the Earth.

The rocket's speed will constantly change as it flies in interplanetary space, being at its greatest during take-off, and gradually decreasing as the rocket recedes from the Earth's orbit.

Having approached Mars, the rocket will by-pass it at a certain distance and fly off into outer space. During the flight around Mars the travellers will be able to photograph the entire surface of the planet, owing to its rotation on its axis.

One year after the take-off the space ship will reach the farthest point of its trajectory, at a distance of 2.175 light-years from the Earth. Here its speed will be at its lowest.

After passing this point the space ship will once more approach the Martian orbit at an increased speed. But this time it will not meet the planet. The elliptical trajectory of the flight having closed, the space ship will return in two years to the Earth at the speed at which it took off.

More powerful rockets will be able to land on Phobos and Deimos, the Martian satellites, from which research can be conducted for long periods. Deimos revolves around Mars in a little more than 30 hours, at a distance of 23,000 kilometers, i.e., $\frac{1}{17}$ of the distance between the Moon and the Earth. Phobos is only 9,000 kilometers from the surface of Mars, and revolves around the planet in less than eight hours. The size and mass of these heavenly bodies are very small and their gravitational pulls are negligible; it will therefore be easier to visit the satellites than their primary.

Modern astrophysics provides data which makes it possible to assume that the natural conditions on the Earth are more similar to those of Mars than of any other planet. Prolonged research by Soviet astronomers, headed by G. Tikhov, has enabled them to come to the conclusion that there is vegetation on Mars. It is believed that the Martian atmosphere contains oxygen and is devoid of gases injurious to human life, although it is very thin even on the planet's surface. The astronauts will therefore have to live in airtight compartments in which the pressure and temperature of the air will be regulated, and space suits will have to be put on before leaving the rocket. It is probable that water is also to be found on Mars. The intensity of solar radiations on Mars is one half of that of the Earth, as a result of which the Martian climate is more rigorous.

What trajectories are the most economical for an expedition planning to land on Mars?

The shortest line between two points in space is a straight line.

However, the space ship cannot travel as the crow flies. The Sun's attraction will deflect the rocket from its route in space in the same way as the Earth's gravity bends the trajectory of a stone flung upwards at an angle. It is true, the trajectory of the space ship can be straightened if its motors work continuously, but then this will increase fuel consumption to an enormous extent. Only if the space ship flies along a vertical line, parallel to the Sun's rays, can it avoid being deflected from its straight trajectory by the Sun's attraction. But a flight of this kind would involve the use of tremendous amounts of fuel, because the space ship would have to neutralize the immense speed of thirty kilometers per second at which it was revolving, together with the Earth, around the Sun. This speed would deflect the space ship off its route in the same way as the current carries away a boat that crosses a river at right angles to the bank.

Let us assume, however, that a journey to Mars is launched along the shortest and most direct route. In that case it would be completed in 85 days, but a speed of not less than 39 kilometers per second would have to be attained for this purpose. It is obvious that such a route would be extremely uneconomical.

On the other hand, a space ship flying along a semi-elliptical trajectory would have to take off from the Earth at a minimum speed. When coming in for a landing on Mars its speed would also be its lowest (Fig. 23).

It has been pointed out that the space ship cannot take off from the Earth at any moment unless it follows a straight course. If the rocket is to meet Mars when it reaches its orbit, Mars must be in a particular position in relation to the Earth, a position that occurs once every 780 days on the average.

A journey to Mars along a semi-elliptical trajectory would take 259 days. Before starting on their way back along the same trajectory the space-travellers would have to wait 454 days for the two planets to stand in the correct relation to each other.

A space ship flying to Mars on such a trajectory would have to attain a speed of 11.6 kilometers per second at take-off. However, it is doubtful whether would-be space-travellers would choose to fly on such a long route. In order to cut transit time they will probably increase take-off speed and fly along a parabolic trajectory. In that case their journey would last 70 days, provided the space ship had an initial speed of 16.7 kilometers per second. Consequently, if the initial speed is increased by 1.4 times the duration of their voyage will be reduced by a factor of 3.7. That is a remarkable feature of navigation in space.

At the end of the last century it was generally believed that Mars was inhabited by higher animals and many stories and novels were written on the subject. The authors never put their characters to

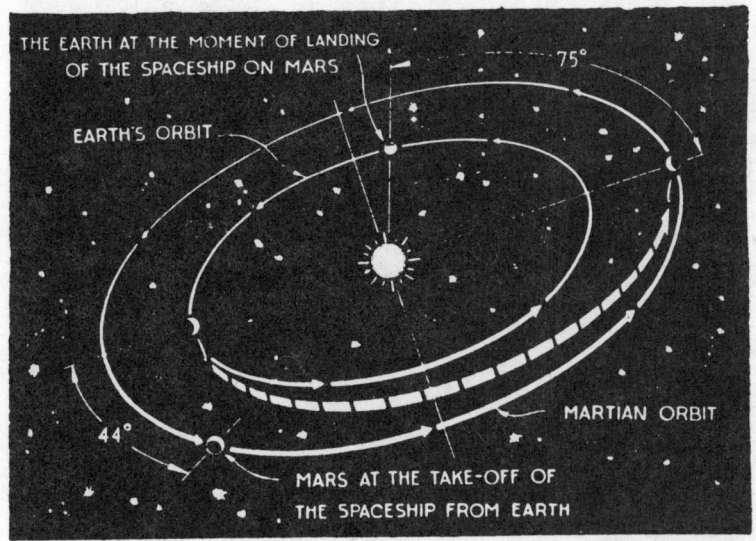

Fig. 23 Flight to Mars along a semi-elliptical trajectory

the trouble of thinking about the timing or the trajectory of their flight. When it comes to the point, however, the situation is much more difficult. A trip from one planet to another can only be made along a number of "reasonable" routes, and the positions of the planets in relation to one another will also have to be taken into consideration. Hence the dates on which space ships might take off or arrive at their destination will be strictly defined.

If a schedule were to be drawn up of flights to Mars or Venus it would have blanks, "dead seasons" of from a few months to one and a half years or so, during which no space ship would be able to take off from the Earth or land at its destination because of unfavourable planetary configuration.

A Trip to Venus

If you look at the darkening horizon immediately after sunset you may often see a very bright star, Venus. Once in a while it makes its apearance before dawn and may even be visible in broad daylight. Venus' brightness is due to its proximity to the Sun and its high power of reflection.

Venus, the Earth's closest neighbour, is more similar to it than any of the other planets of our solar system, its dimensions and mass being only slightly less than those of our planet. Future space-explorers will not therefore be embarrassed by their weight when they land on its surface.

In 1761, Mikhail Lomonosov, with the aid of telescope, discovered a luminous rim around Venus during its approach to the Sun's disc. He explained this phenomenon by the existence of an atmosphere around Venus. Subsequent observations proved that the luminous halo was really the planet's atmosphere, lit up by the Sun. It was observed in 1882 and will not be seen again until 2004. Scientists aboard space ships will be able to see it several times a year.

It was a long-standing opinion that the clouds on Venus were made up of water vapour, which reflect the rays of the Sun fairly well. However, subsequent investigations of the upper strata of the atmosphere have shown that they contain neither water vapour nor oxygen but instead have a high carbonic acid content. It is also probable that the air immediately over the planet's surface is unfit for breathing and the travellers will therefore have to take a reserve of oxygen with them.

While a number of astronomers believe that the structure of Venus' atmosphere is similar to that of the Earth, others consider that it extends to a greater height than that of the Earth. Observations made during twilight on Venus show that the atmospheric pressure on Venus' surface is two to three times greater than on the Earth. This will help to slow down the space ship when it comes in for a landing on the planet's surface.

The period of rotation of Venus on its axis (i.e., the time of one full revolution) has not yet been finally established. Some research workers place it at 68 hours, others take it as equal to the period of rotation of the Earth on its axis, still others consider it to be the same as Venus' period of revolution around the Sun, i.e., 225 days. The angle between Venus' equator and its orbit, which determines the duration of day and night throughout the year, is still not known. And the answers to these questions are not likely to be found until space-travellers fly around Venus. With this data at our disposal, we shall be able to calculate at what altitude and in what direction space ships will have to plunge into Venus' atmosphere to make a safe landing. The less the speed of the space ship in relation to the

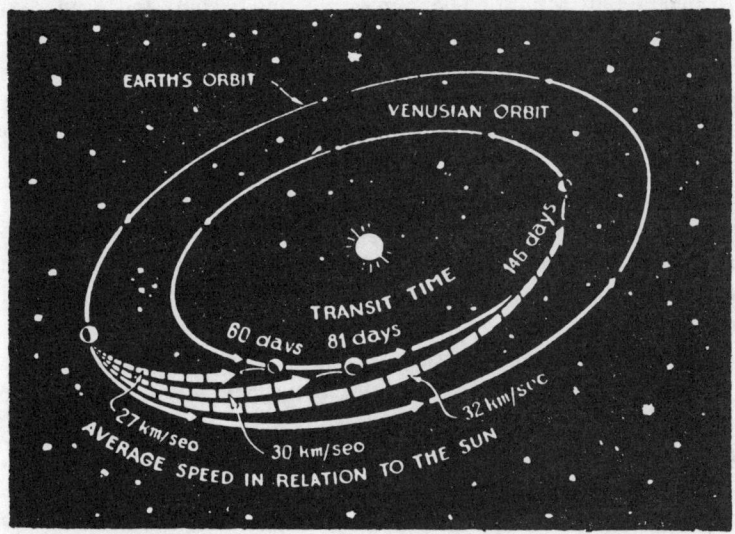

Fig. 24 Flight to Venus along a semi-elliptical trajectory

gaseous envelope of the planet the easier and safer the landing. The speed of the rocket depends to a great extent on whether the direction of its flight coincides with that of the planet's spin.

Initial reconnaissance expeditions will make a thorough study of the structure of Venus' crust and find out whether plants and animals exist there. The cloudy mantle enveloping the planet prevents direct observation of its surface, but new methods of photography by infra-red rays will make it possible to photograph Venus' surface from the space ship in spite of the clouds.

Let us imagine ourselves flying in a space ship to Venus (Fig. 24). After taking off from the Earth at 11.5 kilometers per second the pilot cuts out the motors and the rocket coasts through space like a stone hurled by a sling. The passengers no longer feel their weight and make for the windows, through which they can see our planet, a greenish-blue sphere slowly revolving in pitch dark space a very short distance away. The outlines of the continents, lit by sunlight, are distinctly visible through gaps in the clouds. The ship has escaped from the Earth's gravitational field and is getting farther and farther away from the Earth.

A few months have passed. The Earth has long since become a small bright bluish body. The Sun's heat has become more intense. A new unknown world, glowing bluish-white, is rapidly approaching. It is Venus. The planet grows in size, shutting out more and more stars from view. Now the space ship must be slowed down to prevent its crashing into Venus' surface like a gigantic meteor. If this were to happen, the energy of motion would be transformed into thermal energy in an explosion which would volatilize the metal and everything else, leaving no trace of the space ship except a gigantic crater.

But the pilot has done his best to avoid a crash-landing. He enters the atmosphere of Venus almost parallel to its surface and slows down the rocket by making use of air resistance. Then he switches on the retarding rockets in the nose and practically stops the space ship. A few moments later the space ship makes a safe landing.

The scientists spend their time making observations, experiments, collecting interesting specimens and conducting other research. Finally the date fixed for departure arrives. The space ship takes off at a speed of 10.7 kilometers per second and flies along a semielliptical path at a tangent to the orbits of Venus and the Earth. It will enter the Earth's atmosphere at a speed of 11.5 kilometers per second. To neutralize this speed before landing, the space ship will glide first in the upper layers of the Earth's atmosphere and then in the lower and denser ones.

It will take 146 days to complete such a flight to Venus. However, transit time could be cut down to 81 or 60 days or even less (see Fig. 24).

To attain this goal under terrestrial conditions we would have to increase the speed, as in the case of a stone. The greater the speed with which a stone hurtles through the air the quicker it hits the target. In interplanetary space, however, things are not always like that. In the case described above, the greater the initial speed of the space ship in relation to the Earth, the lower its speed in relation to the Sun, because it will take off in a direction opposite to the Earth's motion. This can also be illustrated by the following example: The faster a man walks in a train in the direction opposite to that in which the train is traveling the slower his speed in relation to the Earth.

If you want to know why the duration of the flight is reduced in spite of the lesser velocity of the rocket's motion in space, have a look at Fig. 24, which gives the clue to the problem. With the lowest of the three speeds shown in this figure the rocket will traverse the shortest distance, which will make it possible to cut travelling time.

Flights to Other Planets

We have described the conditions of travel to the Earth's nearest neighbours—the Moon, Venus and Mars. Flights to other planets of the solar system will be more difficult.

We have seen that the initial speed of flight from the Earth to other planets will depend on the route to be taken and that the semi-elliptical trajectory is the most economical from this point of view. What then are the minimum speeds that we need to reach the other planets of the solar system and how long will such journeys last? Answers to these questions are given in Table II.

From this table we can see that a flight to Mercury along a semi-elliptical path will take less time than one to Venus, although the latter comes nearer to the Earth. The explanation of this phenomenon, however paradoxical at first glance, is given in Fig. 25, which shows that the Earth-Mercury route is shorter than the Earth-Venus route.

Jupiter is several times farther away from the Earth than Mars, and between Mars and Jupiter there is a belt of innumerable small asteroids dangerous to a space ship. The Sun's rays hardly reach this planet. Moreover, parabolic speed on Jupiter is more than five times greater than on the Earth and the force of gravity nearly three times as great, which will be a great hindrance to astronauts and perhaps make it utterly impossible for them to stay on that planet. There are also some other obstacles such as poisonous gases and the cold.

TABLE II.

Planet towards which the space ship is flying	Minimum initial speed km/sec	Transit time (one way)		Initial speed km/sec	Transit time (one way)	
		Years	Days		Years	Days
Mercury	13.5	—	105	—	—	—
Venus	11.5	—	146	—	—	—
Mars	11.6	—	259	16.7	—	70
Jupiter	14.2	2	267	16.7	1	39
Saturn	15.2	6	18	16.7	2	194
Uranus	15.9	16	14	16.7	6	282
Neptune	16.2	30	225	16.7	12	343
Pluto	16.3	45	149	16.7	19	91

However, Jupiter could be investigated from a space ship which circled round it like an artificial satellite.

Before leaving for Mercury one would have to make allowance for the fact that the period of one full revolution of Mercury around the Sun is the same as the period of its rotation about its axis (88 days). Consequently, one hemisphere of this planet is always exposed to the Sun's rays while eternal darkness reigns supreme on the other, owing to which its temperature is very low. The two hemispheres

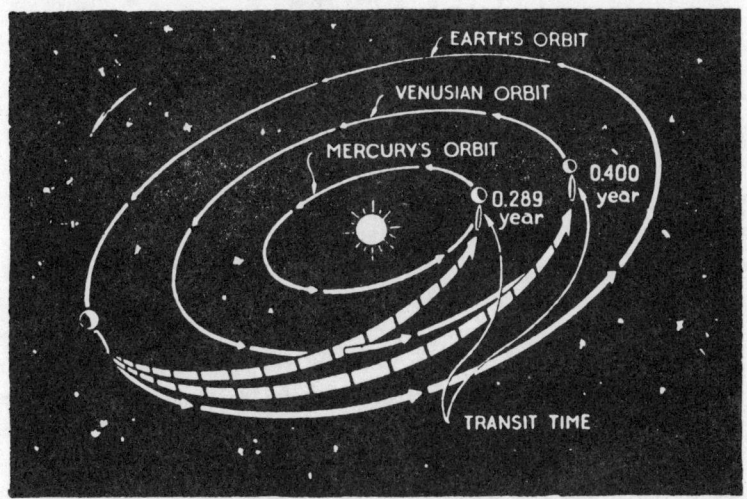

Fig. 25 A flight to Venus along a semi-elliptical path will take more time than a flight to the more remote Mercury.

are divided by a narrow twilight belt with a moderate climate. One must bear in mind, however, that we can speak about Mercury's climate only in a figurative sense, since this planet has apparently no atmosphere at all.

On the average the Sun's rays on Mercury are seven times as powerful as on the Earth. Surface temperature on the Sun-lit hemisphere is as high as 400°C. The skin of a space ship travelling to this planet will have to be designed to reflect most of the Sun's rays falling on it.

A landing on Mercury could only be carried out by rocket-braking, which would be a great handicap.

Flights to Saturn, Uranus, Neptune and Pluto along paths requiring a minimum initial velocity would take too much time. Super-powerful "express" rockets would therefore have to be built to reach them. For example, if the velocity of projection of a Pluto-bound rocket were to be increased by five per cent to the velocity of escape from the solar system (16.7 kilometers per second), the duration of the voyage would decrease to less than one half of the normal. The flight trajectory would then represent the arc of a parabola tangential at the zenith to the orbit of the earth, with a focus in the center of the Sun. The last columns of Table II show the one-way transit times of a flight over such a trajectory to the planets more distant from the Sun than is the Earth.

Although the gravitational pulls of these planets are almost the same as that of the Earth, their natural conditions are unsuitable for man. It has been established that their atmospheres consist chiefly of methane and that their surface temperatures are very low.

What about journeys to the nearest stars?

If we scan the canopy of the heavens with the naked eye we cannot tell the difference between the planets and the stars. But although the two seem to be equally removed from the Earth, the distance between the planets and stars is really very great. Light rays from Pluto, the most distant planet of the solar system, reach the Earth in less than seven hours (the velocity of light is 300,000 kilometers per second, whereas it takes more than four years for light to cover the distance between the nearest visible star and the Earth. Flights to the stars therefore seem to be a matter of the very remote future.

CHAPTER SEVEN

Artificial Satellites of Solar System Bodies

Artificial Satellites of the Moon

The first artificial satellites were created in connection with the program of the International Geophysical Year for exploration of the Earth and the space surrounding it. Now if satellites are good for study of the planet on which we ourselves live, they may prove no less useful for exploring such remote heavenly bodies as the Moon, the Sun and other planets. Artificial satellites can also serve in the capacity of interplanetary space stations. Mastery of interplanetary space will thus require that artificial satellites be created which circle not only around the Earth, but also around the Moon and the planets.

Rockets launched from an interplanetary station or from an orbital space ship can become artificial satellites of the Moon and be used for detailed exploration of the lunar landscape. A Moon satellite circling in an orbit that passes over the lunar poles would be very convenient. The satellite's plane of travel does not, as we are aware, change its orientation relative to stars. This circumstance affords the opportunity to photograph from a satellite the entire surface of the Moon, lighted by sunshine, in four weeks' time, the period of the Moon's rotation around its own axis. But that hemisphere of the Moon, which is lighted by solar rays reflected by the Earth, which is called ash light, will also be clearly visible from on board the cosmic ship since for a lunar satellite this light, with greenish-grey hue, will be ten times brighter than moonlight on Earth in the full Moon period. If this lighting be used as well, then the entire surface of the Moon can be photographed in the course of but two weeks.

In order that the artificial satellite's travel not be disturbed by the Sun and planets, it must circle in the immediate vicinity of the Moon, where the Moon's field of gravitation alone has practical effect.

The circuit period around the Moon of a satellite in orbit at 30 kilometers altitude will amount to 1 hour, 51 minutes and 13 seconds; at 100 kilometers altitude the circuit time is increased by 6.5 minutes. Orbital speeds of artificial satellites of the Moon flying at altitudes reaching several hunderd kilometers will be almost five times less than the speeds of artificial satellites flying at the same altitude above the surface of the Earth. This circumstance considerably facilitates observation of the Moon's surface. From an altitude of 30 kilometers it will be possible even with the naked eye to distinguish objects 3.8 to 8.7 meters in size, (depending on eyesight keenness) that are on the Moon's landscape. But at the travel speed corresponding to this altitude—upwards of 1600 meters a second—and with the comparatively small diameter of the field of vision—about 650 kilometers—the object under observation will rapidly disappear from view. Therefore, a satellite which circles at a greater distance, for example 150 kilometers from the Moon, will be more convenient for observation. Then at somewhat lower travelling speed the diameter of the lunar area embraced by the eye would be considerably greater, reaching 1400 kilometers. Objects on the Moon's surface would remain longer in the observer's field of vision. In this case, however, it would be possible with the naked eye to observe only objects with a diameter from 19 to 44 meters. We note, by the way, that inasmuch as expeditions of this kind will have powerful optical units, even the finest details will be observable. The maximal visibility time of any point on the lunar surface from a satellite space ship which is at an altitude of 30 kilometers will amount approximately to 6.4 minutes, and at an altitude of 150 kilometers to 15.7 minutes.

If close-up study of any detail on the lunar landscape is required, then at the price of a small expenditure of fuel the artificial satellite travelling in a circular orbit at an altitude of 30 kilometers can switch over to an elliptical orbit with its periastron located at the desired altitude. (Periastron is defined as that point in the orbit of a heavenly body, as of a satellite, which is nearest to the center around which it revolves.) To lower the periastron right down to the very surface of the Moon, it is necessary to reduce the speed by just 7.14 meters a second. But only after 54 minutes 3 seconds, having beforehand completely circled one hemisphere of the Moon, will the artificial satellite approach the locality marked for study. The satellite will then automatically return to the former flight altitude of 30 kilometers, where an increase of speed by the same 7.14 meters a second will lead to the satellite switching over to the former circular orbit. In general the switching of the artificial Moon satellite to another higher or lower circular orbit requires very small changes of speed, and consequently, small expenditure of fuel. For instance, to switch over to zero circular orbit at the moment of passage through the periastron it is necessary to reduce speed by just 7 meters a second.

When the satellite gets into the shadow of the Moon, night falls on it, which as in the case of the artificial Earth satellite will always be shorter than the local day. The day-night cycle on a satellite which is at an altitude of 122 kilometers above the Moon will be just 2 hours long. Every 10 kilometers of increase in altitude will cause a lengthening of the cycle by approximately 1 minute, and in reduction of altitude, the same proportional shortening of the day-night cycle.

In order to speed up the survey of the entire surface of the Moon, it will not be necessary to wait passively until the Moon has made a half-turn or a complete revolution around its axis. After making one circuit around the Moon with switched-off engine, and surveying a zone of definite width, the space navigators can while flying over the pole switch on the rocket engine in order to turn the plane of the satellite orbit around the axis of the Moon at a certain angle with respect to the stars. In such a manner the astronauts will see completely new territories on the satellite's next circuit in a new orbit around the Moon. The higher the altitude of satellite travel, the wider the visible zone and the less frequently the plane of the satellite orbit has to be changed. So, with flight altitude of 10 kilometers this orbit shifting has to be done 14 times, but with flight altitude of 50 kilometers, only 6 times (see Table III). Employing this method, it will be possible instead of waiting passively for two weeks, to survey the entire surface of the Moon in 1 day, 3 hours and 20 minutes from an altitude of 10 kilometers, and almost twice as fast from an altitude of 50 kilometers.

Naturally this will not be inexpensive: in the first case for every turn of the orbit's plane as much fuel would have to be expended as is required to give the rocket a speed of 350 meters per second, and in the second case a speed of 736 meters per second. Besides, the rocket engine's thrust will each time be so directed that only the orientation of the orbit's plane be changed while the orbital velocity of the Moon satellite remained the same.

With flight around the Moon by such a method it will be possible to survey some part of its surface even twice, three times and so forth. Illustrated in Fig. 26 is the method described of a speeded-up survey of the Moon from its artificial satellite flying at an altitude equal to the radius of the Moon. In this case it will suffice to change the plane of the artificial satellite's orbit only twice. The figures in the drawing indicate how many times a given surface can be surveyed from the satellite with a single flight around each orbit.

At the astronautical congress in Rome in 1957, the opinion was expressed that artificial satellites of the Moon will be created before a manned artificial satellite of the Earth can be launched.

Artificial Satellites of Planets and the Sun

Artificial satellites circling around other bodies of the solar system will, as already mentioned, have great importance for interplanetary flights. The orbital velocities and circuit periods of these satellites will vary within very wide limits. And these magnitudes characterize on the one hand the degree of difficulty in launching them, and on the other the degree of their usefullness in the capacity of observation station.

Let us compare the circular velocity and sidereal circuit period

TABLE III.
Characteristics of Two Artificial Satellites of the Moon

Characteristics	Flight Altitude above Moon Surface in kilometers	
	10	50
1. Circular velocity in meters per second	1674	1655
2. Its ratio to zero circular velocity, in per cent	99.71	98.59
3. Increase of circular velocity with reduction of flight altitude by 1 kilometer, in meters per second	0.479	0.463
4. Radius of orbit in kilometers	1748	1788
5. Ratio of radius of orbit to radius of Moon, in per cent	100.6	102.9
6. Length of orbit in kilometers	10,983	11,234
7. Angular velocity in angular seconds per second	198	191
8. Ratio of angular velocity to zero angular velocity, in per cent	99.1	95.8
9. Circuit period	1 hr. 49 min. 20 sec.	1 hr. 53 min. 7 sec.
10. Minimal length of day	58 min. 24 sec.	1 hr. 5 min. 5 sec.
11. Ratio of day length to circuit period, in per cent	53.4	57.5
12. Maximal length of night	50 min. 56 sec.	48 min. 1 sec.
13. Ratio of maximal night length to circuit period, in per cent	46.6	42.4
14. Minimal arc of orbit through which satellite passes in the shadow of the Moon	192° 16′	207° 10′

TABLE III. *(Contd.)*
Characteristics of Two Artificial Satellites of the Moon

Characteristics	Flight Altitude above Moon Surface in kilometers	
	10	50
15. Maximal arc of orbit through which satellite passes in solar rays	167° 44′	152° 50′
16. Arc described by the satellite in the sky from the viewpoint of an observer who is on the Moon in the plane of the orbit (center of the arc in the center of the orbit)	12° 16′	27° 10′
17. Length of the arc of the global segment of Moon (along the great circle) visible from the artificial satellite, in kilometers	372	824
18. Ratio of area of visible global segment to surface of the Moon, in per cent	0.286	1.398
19. Maximal duration of observation of a point on the surface of the Moon	3 min. 42 sec.	8 min. 32 sec.
20. Minimal linear dimensions of objects distinguishable in binoculars with 15-power magnification, in meters:		
(a) with average vision	2.91	14.54
(b) with keen vision	2.26	6.30
21. Free fall acceleration in artificial satellite orbit in meters a second per second	1.60	1.53
22. Ratio of preceding magnitude to free fall acceleration on the surface of the Moon, in per cent	98.9	94.5

TABLE III. (*Contd.*)
Characteristics of Two Artificial Satellites of the Moon

Characteristics	Flight Altitude above Moon Surface in kilometers	
	10	50
23. Minimal number of turns of the orbit's plane necessary for complete survey of the Moon	15	7
24. The angle between consecutive positions of the orbit's plane	12° 0'	25° 42'
25. Minimal arc of overlapping of observations	0° 16'	1° 28'
26. Velocity necessary for a turn of the orbital plane, in meters per second	350	736
27. Total velocity necessary for completing all turns of the orbital plane in meters per second	4900	4418
28. Minimal duration of survey of entire Moon surface	27 hrs. 20 min.	14 hrs. 57 min.
29. Ratio of surveyable surface to Moon surface, in per cent	160	169
30. Velocity of take-off from artificial Moon satellite to the Moon, in meters per second	3	12
31. Velocity of landing on the Moon, in meters per second	1681	1691
32. Summary velocity in descent to the Moon from an artificial satellite, in meters per second	1683	1703
33. Duration of the flight from the satellite to the Moon	53 min. 30 sec.	54 min. 31 sec.

of a zero artificial satellite of the Earth with analogous data for zero satellites of other planets and the Sun.

If identical force of attraction prevails on the surfaces of different celestial bodies (which is approximately true, for example, for the Earth, Uranus and Neptune, and also for the second and third satellites of Jupiter—Europa and Ganymede), then the circular velocity will be the greater on that one of the examined bodies whose di-

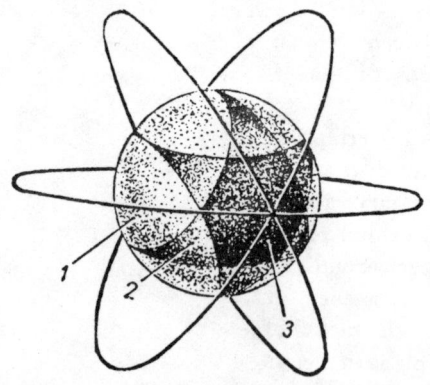

Fig. 26 Speed-up method of surveying the Moon's surface. After completing a circuit around the Moon, the space travellers modify over the pole the plane of the satellite's rotation orbit. The figures indicate how many times a given area can be surveyed from the satellite in case of a two-fold change of the plane of the orbit, situated at an altitude equal to the radius of the Moon.

mensions are greater. On the other hand, in the case when two celestial bodies have identical sizes (for example, approximately the Earth and Venus, or the Moon and the first satellite of Jupiter—Io), the circular velocity will be the greater on that body where the force of attraction is greater.

The zero circular velocity for the Earth is, as is evident from Table IV, greater than for the three remaining inferior planets (Mercury, Venus and Mars). But it is one-half that for Uranus and Neptune, and one-fifth that for Jupiter. For the Sun the zero angular velocity is 55 times greater than the zero circular velocity for the Earth and upwards of 10 times greater than for Jupiter.

The sidereal circuit period of a zero artificial satellite of the Earth

TABLE IV.
Zero Circular and Parabolic Velocities for Planets and the Sun

Name of Celestial Body	Zero Circular Velocity in km/sec	Zero Parabolic Velocity in km/sec	Ratio of Zero Circular and Parabolic Velocities to Corresponding Velocities for Earth
SUN	437,535	618,753	55.300
MERCURY	3,038	4,282	0.383
VENUS	7,319	10,351	0.925
EARTH	7,912	11,189	1.000
MARS	3,562	5,038	0.450
JUPITER	42,205	59,686	5.334
SATURN	25,100	35,495	3.172
URANUS	15,308	21,648	1.935
NEPTUNE	15,129	22,810	2.039

TABLE V.
Sidereal (Star) Circuit Periods of Zero Artificial Satellites of Planets and the Sun

Name of Celestial Body	Sidereal Circuit Period of a Zero Satellite		Ratio of This Period to Sidereal Period of Zero Satellite of the Earth
	Hours	Minutes	
SUN	2	46	1.98
MERCURY	1	24	1.00
VENUS	1	27	1.04
EARTH	1	24	1.00
MARS	1	40	1.19
JUPITER	2	57	2.11
SATURN	3	45	2.68
URANUS	2	50	1.90
NEPTUNE	2	62	2.02

together with that corresponding for Mercury, are shortest among all bodies of the solar system (Table V). For Venus and Mars this period is somewhat greater than for the Earth, and for the Sun,

Uranus, Neptune and Jupiter it is twice greater. A zero artificial satellite of Saturn has the longest sidereal circuit period (three times greater than for the Earth).

What is the explanation of the differing magnitudes of the duration of the sidereal circuit period for the various planets? It can be shown that the circuit period of an artificial satellite at the surface of a celestial body depends exclusively on the average density of a given celestial body. The less the density, the longer the circuit period which is inversely proportional to the density's square root. If the density of a heavenly body is, for example, four times less than the density of the Earth (the Sun, Jupiter, Uranus and Neptune have approximately such density), then the circuit period of the satellite at the surface will be twice longer. With ninefold increased density, the circuit period is reduced three times, etc.

The planets of the solar system can, from the viewpoint of average density, be divided in the main into two groups, in each of which the density is rather close. To the first group belong Mercury, Venus, Earth, Mars and apparently Pluto; to the second belong the Sun, Jupiter, Saturn, Uranus and Neptune, the density of Saturn being one-half that of the Sun. The zero circuit period for Saturn is the longest in the solar system.

Thus, the circuit periods of zero artificial satellites for Mercury and Earth, that is planets having approximately equal density, are almost the same, despite the fact that the diameter of Mercury is 2.63 times less than the diameter of the Earth. But the circuit period of an artificial satellite at the surface of the Moon, which has less density than the Earth, will be longer than the circuit period of a zero satellite of our planet, despite the lesser dimensions of the Moon.

To the present day, we note, the question about the density of Mercury has not been clarified. According to the American astronomer Simon Newcomb (1835-1909) the density of Mercury is approximately equal to the density of the Earth. But according to the estimates of the Russian astronomer O. A. Baklund (1846-1916), it is one and a half times less. This question might be solved, if an artificial satellite of Mercury were created. By the period of its circuit the planet's density could be established. (Mercury does not have its own natural satellite.)

The zero artificial satellite of the Sun (it is better to say "zero artificial planet") must, as is evident from Table V, complete a full circuit around the day luminary in 2 hours and 46 minutes. In other words, such would be the length of a year on it. In case the artificial planet has an orbit radius equal to 100 astronomical units, the duration of a year on it would amount to a thousand years, and with a radius ten times larger still a year's duration would amount to 31,623 of our years.

(The average distance of the Earth from the Sun, equal to 149,500,000 kilometers, is taken to be an astronomical unit. Giant stars are known to exist which consist of exceptionally rarefied substance; their density amounts to from one thousandth to one hundred-millionth fraction of the density of the Sun. The circuit period of satellites at the surfaces of these stars would be from 4 days to upwards of 3 years. The so-called white dwarfs, on the contrary, have a density which is 7000 and more times greater than the density of the Earth. The circuit period of a zero artificial satellite around such a star would be just 1 minute.)

The characteristics of satellite space stations of planets and the Sun are given in Table VI.

TABLE VI.
Characteristics of Satellite Space Stations of Planets and the Sun

Celestial Body	Stellar Days* and Sidereal Circuit Period of Satellite			Radius of Satellite Orbit in Body Radii	Radius of Satellite Orbit in Thousands of Kilometers
	Days	Hours	Minutes		
SUN	25.38			36.45	25,350
MERCURY	87.97			131.5	318.4
VENUS		68		2.802	17.08
EARTH		23	56	6.614	42.19
MARS		24	37	6.020	20.42
JUPITER		9	50	2.231	159.2
SATURN		10	14	1.952	117.9
URANUS		10	42	2.425	60.25
NEPTUNE		15		3.014	79.87

* The stellar days indicated for Jupiter and Saturn are for the equator, since the surface of each of these planets participates in the daily rotation not as one whole.

Artificial satellites of planets like Earth satellites actually cannot circle in an orbit at any distance from their central heavenly body. At a very great distance from the planet they may be captured by the Sun's powerful field of gravitation. Existing natural satellites can also more or less distort the trajectories of artificial satellites, and in exceptional cases even capture them.

The radii of the spheres of attraction of planets according to Francois Tisserand (1845-1896) are cited in Table VII. It strikes the eye at once that Mars having a mass one half the mass of Venus has the same sphere of attraction as Venus, and Uranus whose mass

TABLE VII.

Spheres of Attraction of Planets (According to Tisserand)

Planet	Radius of sphere of attraction in astronomical units	Radius of sphere of attraction in millions of kilometers
MERCURY	0.001	0.15
VENUS	0.004	0.6
EARTH	0.006	0.9
MARS	0.004	0.6
JUPITER	0.322	48.1
SATURN	0.363	54.3
URANUS	0.339	50.7
NEPTUNE	0.576	86.1

is almost 22 times less than the mass of Jupiter, nevertheless has a sphere of attraction somewhat greater than Jupiter's sphere of attraction. This is explained by the fact that Mars is farther from the Sun than Venus, and Uranus is farther than Jupiter, and therefore the Sun's sphere of influence near Mars or Uranus is considerably less than near Venus or Jupiter.

As for the launching of artificial planets, modern scientists express preference for launching in a semi-elliptical trajectory. In this connection we direct the reader's attention to certain characteristics of such launching of artificial planets.

We saw above that for the launching of an artificial satellite in a semi-ellipse at an altitude, for example, of 15 radii of the Earth (counting from its center), less summary velocity is required than

Artificial Satellites of Solar System Bodies

at an altitude several times higher. A similar picture is observed in the launching of artificial planets.

The launching of an artificial planet rotating around the Sun in the orbit of Jupiter will naturally, owing to the remoteness of this orbit, require greater summary velocity than the launching of an artificial planet in an orbit of Mars, adjacent to the Earth. The creation of an artificial satellite which would travel in a circular orbit of greater dimensions than the orbit of Saturn, will however require less summary velocity than the launching of a planet in the orbit of Saturn. More detailed data is given in Table VIII.

Let us imagine that from the Earth in a semi-ellipse is launched a rocket probe which must be converted into an artificial planet

TABLE VIII.

Launching of Artificial Planets in Semi-Elliptical Trajectory

	SATURN	URANUS	NEPTUNE	PLUTO
Average distance from the Sun, in astronomical units	9.54	19.19	30.07	39.46
Velocity of take-off from surface of Earth, in kilometers per second	15,213	15,897	16,164	16,279
Additional velocity communicated to the Rocket at the moment of arrival in orbit, in kilometers per second	5,441	4,661	4,054	3,685
Summary Velocity, communicated to the rocket for its conversion into an artificial planet, in kilometers per second	20,654	20,558	20,218	19,964
How much less the summary velocity is than the corresponding velocity for the orbit of Saturn, in kilometers per second	—	0.096	0.436	0.690

rotating around the Sun in the orbit of Saturn. However, owing to an accident at take-off, a fuel leak occurred and velocity equal to 400 meters per second was lost. Then, acting on a radio command from Earth, pumps began to supply to the combustion chamber fuel which was previously intended for additional increase of speed at the moment the rocket reached a point located the distance of Saturn from the Sun. In consequence not only was the loss of velocity compensated, but instead of the projected 15.21 kilometers a second speed of take-off from Earth, the rocket developed a speed that was 0.95 kilometers a second greater. Owing to this it intersected the orbit of Uranus which is travelling at a distance from the Sun twice greater than the distance of Saturn, and reached the orbit of Neptune, the radius of which is more than three times greater than the radius of the orbit of Saturn. At that moment the rocket engine was again switched on automatically. Almost all the remaining fuel which was intended for the additional thrust in switching over to the orbit of Saturn was expended, and the rocket was converted to an artificial planet travelling in the orbit of Neptune. For launching an artificial planet in the orbit of remote Neptune less fuel was necessary, as we see, than for creation of an artificial planet travelling in the orbit of Saturn which is situated considerably nearer.

Artificial Satellites of Comets

Artificial satellites equipped for exploration of comets are obviously of considerable interest.

Comets usually consist of a head, which is a nucleus surrounded by a nebulous envelope, and a long tail. As an exception comets are encountered having several heads or several tails. The nucleus is a cluster of solid bodies more or less large in size, surrounded by an envelope of dust, having considerably greater density than the substance of the tail.

In degree with approach to the Sun the comet head is compressed, and its gaseous tail is elongated. The latter is a substance so rarefied that a cosmic ship, flying through it, would not experience any resistance. "A visible nothing", "a bag of emptiness", that is what astronomers say of the tail.

Comets are extremely "transient" bodies. During every approach to the Sun, an enormous quantity of the gas forming the tail is evaporated from the head parts. This gas is dispersed afterward into space; on every subsequent approach to the Sun, therefore, the tail of the comet becomes less and less lustrous; then it disappears entirely. All comets whose bright and luxurious tails frequently shine in the black velvet of the heavens were born in recent centuries and milleniums.

In the formation, life, and death of comets are many obscure problems which might be solved by close-up observation of a comet if only during one of its complete circuits around the Sun. That such an expedition will sometime be carried out by astronauts is indisputable.

Comets can be observed from the Earth only when they are close to the Sun. But space travellers will be able to observe all the changes which occur in the comet, even during its maximal recession from the Sun, and can trace all stages of the modification of its head and tail forms.

It is still early at present to speak about the landing of space explorers on the large mass lumps which can be found in the comet's head part. The very approach to the comet's head part may not be without peril for the space ship: here the probability of collision with meteorites is especially great. One can therefore speak only about travelling along with the comet at a safe distance away. After equalizing the speeds of ship and comet, and then switching off the rocket engine, the space explorers will escort the heavenly body at a distance convenient for observation.

But it would be most convenient to study the comet's nucleus through converting the rocket into its artificial satellite. Owing to the comet's small force of gravitational attraction, such a satellite would circle in orbit around the nucleus with extremely low speed, which would facilitate observations. The artificial satellite might travel into the "very thick" of the substance of the comet's head, without meeting practically any kind of medium resistance (if one omits the already mentioned threat of collision with meteorites).

The comet observed in 1818 proved to be the largest in size and mass known to us. Its nucleus had a diameter of 20 kilometers, and the mass amounted to 2.10^{13} tons. An artificial satellite travelling at the "surface" of this comet's nucleus would possess a speed of 10 meters a second (approximately the speed of a passenger train). Under such conditions the survey of the comet during one of the ship's circuits around the nucleus would occupy just 1 hour and 45 minutes. (Of course, a definite flight altitude should be selected for photographing the comet's nucleus.)

If it should turn out that the comet's nucleus rotates around its own axis (which is still unknown), then by continuing the flight in the former orbit, it would be possible from the artificial satellite to survey the nucleus from all sides. In the opposite case, by means of the rocket engine the plane of the satellite's orbit might be turned after each circuit of the artificial satellite (or after several circuits). Besides, the same consumption of fuel would be required for all circuits with the object of a complete survey of the comet's nucleus as is needed to communicate to the rocket a speed of 31 meters a second.

For study of the substance of the comet's head, which surrounds the nucleus, the rocket might make a "jump" to a certain height and switch over to an elliptical orbit, so that the circuit period would be, for example, several days or weeks. For this maneuver it would be sufficient to communicate to the rocket an additional speed of less than 5 meters per second.

Orbital Space Ships

Not all artificial satellites will, as has already been said above, circle around the Earth in its immediate vicinity. Theoretical estimates show that it will be possible also to create such motorized artificial satellites of the Earth and artificial planets (i.e. artificial satellites of the Sun), which having an elongated elliptical orbit will with appropriate correction of the orbit by means of the rocket engine (for extinguishing the perturbing effect of other celestial bodies) cruise on regular routes in the universe, serving as space travel lines. They will rotate in their orbits like the planets and their satellites, periodically passing at the same time close to the Earth.

We will call such artificial heavenly bodies orbital space ships, since it would be possible to use them for the purposes of cosmic travel routes. An expedition going from the Earth to the Moon, for example, can use an orbital space ship as a transfer means of travel. Flying up to such a ship on a small rocket, the space travellers can transfer to it and continue the journey further. Then, on approaching the Moon, the travellers again will transfer to a small rocket in order to land on the Moon's surface.

Artificial Satellites of Solar System Bodies

Dwelling quarters, workshops and observatories will be erected on orbital space ships. Here the astronauts will find everything necessary for further flight.

We will first discuss orbital satellite space ships which circle around the Earth and the Moon at the same time. Such ships will travel regularly in orbits passing close to the surfaces of these two celestial bodies. The orbit of the ship can be so calculated that in each sidereal month (the interval of time during which the Moon, circling around the Earth, returns to its former position relative to the sky) the ship will fly over the hemisphere of the Moon that is invisible from the Earth. If for example, the major axis of the orbit will be equal to 484,318 kilometers then the orbital ship will complete

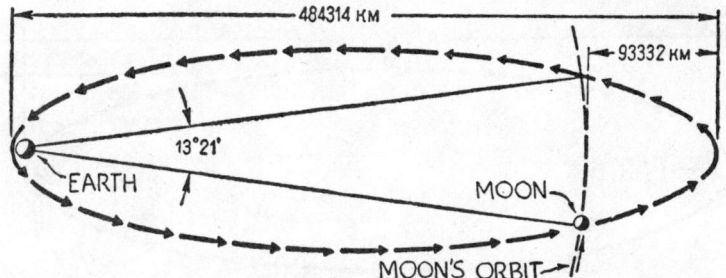

Fig. 27 This trajectory of an orbital space ship will permit observing a wide zone of the Moon hemisphere hidden from our view. The explorers will encounter the Moon after each sidereal month.

two rotations with respect to the stars during the time that the Moon makes one rotation. The trajectory can be planned with the object such that the artificial satellite will intersect the Moon's orbit at the desired distance from its surface. From the Earth or from an interplanetary space station rotating at a low altitude, it will be possible to fly (with a speed somewhat in excess of 3 kilometers a second) at such a moment that the orbital ship will intersect the lunar orbit before the Moon reaches that point. (Here and subsequently, when the point under discussion is a take-off from an interplanetary space station, estimates are made on the assumption that the artificial satellite is travelling in a circular orbit at an altitude of 200 kilometers with a speed equal to 7791 meters per second, corresponding to that altitude.) Observers will thus be able to survey a wide zone close-up of the Moon hemisphere hidden from our eyes (Fig. 27).

If this orbital ship will in its perigee fly at an altitude of 200 kilometers above the Earth's equator, then its flight to the Moon will last 3 days, 3 hours and 20 minutes. After intersecting the lunar orbit the ship recedes from the Moon an additional 93,337 kilometers, and then begins to return to the Earth and after 7 days 9 hours 11 minutes again intersects the orbit of the Moon at a distance of 13° 21' from the first point of intersection. Along an oblate ellipse with a minor axis at 112,120 kilometers, the orbital ship will return to Earth, make one rotation "idle" without meeting the Moon, and after 27 days 7 hours 43 minutes from the take-off the whole cycle is repeated with the sole difference that the phase of the Moon will now vary for the observers. On such a ship the astronauts will in

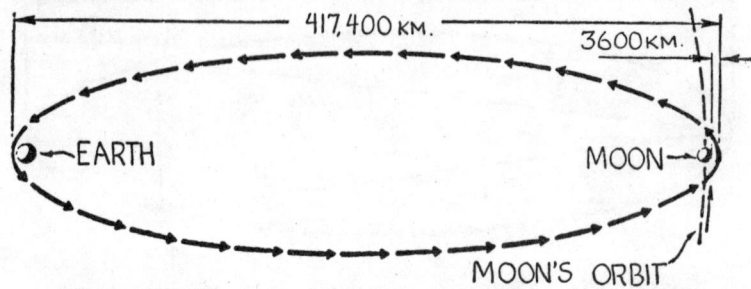

Fig. 28 Along such a trajectory the orbital space ship will pass close to the Moon at a distance of 3600 kilometers. But this will occur only once in two sidereal months; during that time the ship will pass close to the Earth five times.

the course of a year fly thirteen (and sometimes fourteen) times to the Moon which will each time be in a changed phase. Every two weeks the space travellers will have an opportunity to descend from the orbital ship to the Earth. At the same time various cargo, in particular provisions, can be also transferred from the Earth to the ship.

Such an orbital ship, however, has a shortcoming: it recedes too far from the Moon and passes by it at very high speed. Preferable from this view point is an orbital ship travelling in another orbit. It will fly at an altitude of but 3600 kilometers above the Moon during its passage through the point of its orbit farthest removed from the Earth (Fig. 28). But such an orbital ship has another shortcoming: it will fly close to the Moon only once in two months, going around the Earth five times in that same period.

A satellite space ship making three circuits around the Earth in a month can be launched for exploring from a comparatively low altitude the hemisphere of the Moon that is visible from the Earth. From a perigee at an altitude of 200 kilometers above the Earth, such a satellite would in 4 days, 13 hours and 17 minutes fly to an apogee lying at a distance of 363,026 kilometers. Because of the eccentricity of the lunar orbit such an orbital ship might fly very close to the Moon.

The trajectories of "lunar" orbital ships are plotted in Figs. 27 and 28. But that is how they will look when the ship flies through "idle" without meeting the Moon on its path. With approach to the apogee their velocity will gradually decline, falling in the apogee to 150 to 200 meters a second. After passing through the apogee, the ship will again begin to pick up speed. But when after one or several "idle" circuit runs, the orbital ship gets into the Moon's sphere of attraction, then with approach to the apogee its velocity will not decline, but on the contrary, owing to the pull of the Moon, will begin to increase. After passage through the apogee, when the ship again begins to approach the Earth, the Moon's field of gravitation will brake its travel and only at a great distance from the Moon will the Earth's gravitation finally exert its influence, accelerating the satellite's travel. Besides, the elliptical form of the ship's trajectory is naturally distorted considerably.

The Earth's compression, as already mentioned, also has some effect on the satellite travel, owing to which the orbit's perigee and apogee migrate and the orientation of its plane is changed. The so-called disparity of lunar motion (deviation from the laws of Kepler) also exerts a certain influence. As a result of all these causes, an orbital ship which flew past the Moon for many months—even at a great distance from it—might even collide with it after a certain length of time. Constant correction of the trajectories of such satellites, orbital space ships, by means of the rocket engine is therefore quite essential.

Now let us turn to orbital space ships which might travel around the Sun as artificial planets. Their orbits will differ from the orbits of natural planets by reason of their great eccentricity: the distance of the orbital space ships from the Sun will vary sharply, owing to which they can serve as means of travel between the planets.

Let us examine one of the possible alternate trajectories of an orbital space ship flying on the Earth-Venus route (Fig. 29).

An orbital space ship having on board an expedition for exploration of Venus, takes off from the surface of the Earth or from an interplanetary station, with a speed somewhat in excess of 4 kilo-

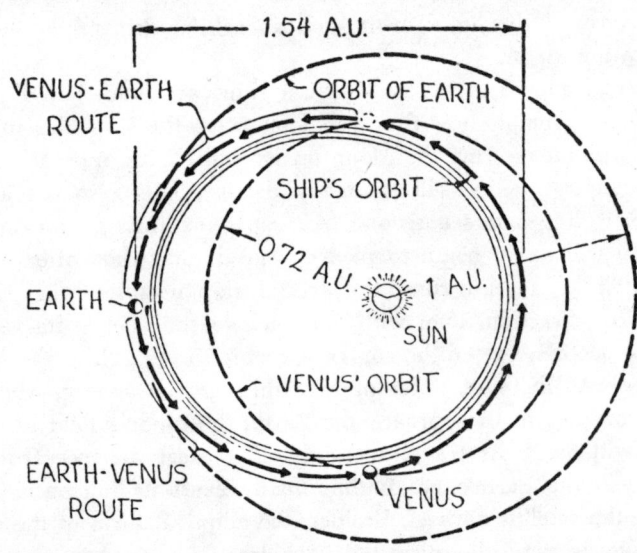

Fig. 29 Orbital space ships can also cruise around the Sun. The diagram shows one of the possible alternate trajectories of an orbital space ship travelling on the Earth-Venus route.

meters a second. It flies in a planned orbit tangent to the Earth's orbit. After 81 days it flies through near Venus. A landing party disembarks to the surface of the planet by means of a cosmic glider, while the ship travelling on in its orbit approaches the Sun at a distance of 0.54 astronomical units and returns to the point of departure in the Earth's orbit 8 months later. During that time the Earth has not yet had time to reach this point. When after another 8 months, the orbital space ship against returns to the take-off point, it will again miss the Earth there, since it had passed through four months earlier. And only after two years from the moment of take-off, having made three complete rotations around the Sun, does the orbital space ship encounter the Earth.

Meanwhile, the expedition landed on Venus is engaged for a year and a half in exploration of the planet and prepares everything to take off at a date calculated beforehand, when the orbital space ship again flies through near Venus. In our case the duration of the expedition's stay on Venus amounts to 2 years minus 162 days, the round-trip time of a flight to and from Earth. So, in 568 days after the descent the landing expedition returns to the orbital space ship flying just then past near Venus, and boarding it is again delivered to Earth. The astronauts descent to the Earth, while the ship continues its perpetual motion in the cosmos.

After some time a second expedition can use the same orbital space ship. But for this purpose it is essential that at the moment of the expedition's take-off from Earth, Venus and the Earth assumed the same mutual position as during the first flight. The moment of take-off from Earth can be calculated with the object of having the ship encounter Venus on the very first intersection of its orbit (which is what we presupposed when we described the first flight).

It is readily understood that the next such take-off to Venus can be made when the integral number of Venus years is equal to the integral number of Earth years. Venus makes a complete rotation around the Sun in 0.6152 Earth year. Therefore, of course, complete matching of the number of rotations of the orbital space ship and the Earth is not to be expected. But if a matching that is not absolutely accurate is satisfactory, then the next flight can be made in 6 years after the return of the Venus expedition, i.e. in 8 years from the moment of this expedition's take-off from Earth. During these 8 terrestrial years Venus will make 13 complete rotations around the Sun and its position relative to the Earth is repeated almost exactly (a year on Venus lasts approximately $8/13$ of an Earth year $- 8/13 = 0.6154$); the discrepancy derived is only approximately by one degree in the arc of the Earth's orbit, which is readily compensated for by appropriate correction of the trajectory. "The schedule of work" of our new expedition will be, of course, just the same as that which the first expedition had.

Here we have told of only some alternate cosmic flights in artificial satellites and artificial planets. But other trajectories are also possible for orbital space ships, from which exploration of the universe will be made. In Fig. 30 are shown the routes of orbital space ships —artificial planets which will return automatically to the Earth in one year (1), two years (15) and three years (4 and 8).

Our calculations show that there are 24 trajectory ellipses for orbital space ship-planets travelling inside the orbit of the Earth all the way to the very surface of the Sun and passing close to our planet at time intervals which are expressed by an integral number of years from 1 to 5. Moreover, an additional 39 orbits exist in this same space, along which the ship will pass by the Earth every 6,

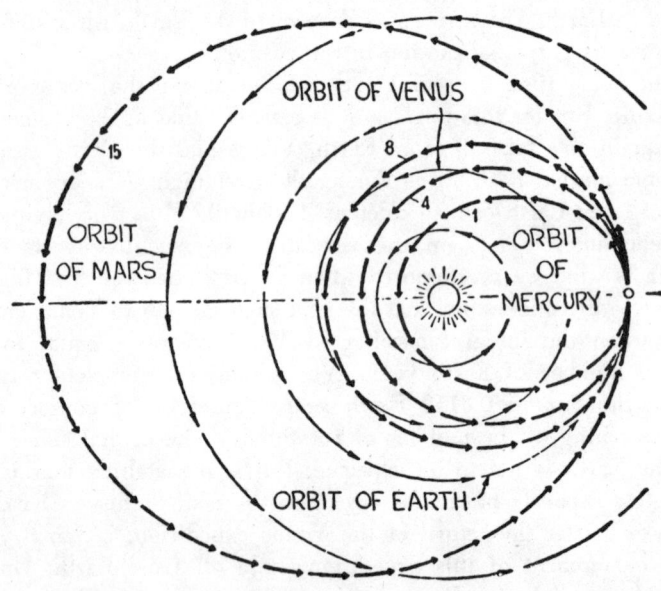

Fig. 30 Routes of orbital artificial planet-space ships which return automatically to the Earth after one year (orbit 1), two years (orbit 15), and three years (orbits 4 and 8). The orbit number corresponds to Table IX.

7, 8, 9, and 10 years. For exploration of the space between the orbits of Earth and Jupiter, there are 27 trajectories which pass outside the orbit of our planet. Travelling along these trajectories, the space explorers can return to Earth after 2, 3, 4, 5 and 6 years.

The characteristics of certain orbital artificial planet-space ships which return to Earth after a time interval of one to five years are given in Table IX. For the orbital ships which rotate inside the orbit of the Earth, we confine ourselves here only to those orbits (numbers 1 to 10) which approach the Sun at a distance not closer than 0.260 astronomical unit.

To Whom Does Outer Space Belong?

The appearance of the first shoots of the newest branch of jurisprudence, space law, is one of the symptoms of the incipient era of cosmic flights. Numerous articles, books and dissertations are devoted to it, and in some foreign countries lectures are given on this theme.

A special course on space law, which is a component part of the study program, is given in the Institute of International Air Law at the University of Montreal (Canada). Jurists there are who attempt in advance to defend the "rights" of certain circles to interplanetary space, to the Moon or a neighboring planet. Others, on the contrary, are guided by considerations of the love of peace and humanity. As early as upwards of a quarter of a century ago, the German author V. Mandl wrote a book under the title of *Das Weltraum-recht* ("Space Law"). At international astronautical congresses, legal problems of space flights have been discussed by the President of the American Rocket Society, A. G. Haley, the deputy secretary general of the United Nations Organization, O. Schachter, and others. In April of 1956, the American international law society devoted a special meeting to problems of space beyond the atmosphere.

Problems of the sovereign rights of various states in the modern definition of these rights are to be encountered already in the testing of ultra-long-range and high altitude rockets.

The development of astronautics will naturally bring to the fore the question of the sovereignty of states with respect to the space located above their territories.

International air law proceeds from the primary position of recognizing the state's full sovereignty with respect to the air space extending above the territory subject to its jurisdiction. Many authors think that while this law cannot serve as the basis of principle for interplanetary space law, it can help in the development of space law principles.

Certain juridical and political complications that might arise with

TABLE IX.

Main Characterics of Orbits of Orbital Artificial Planet-Space Ships Which Return from Outer Space to Earth After Intervals Not over Five Years

Orbit No.	Distance from the Sun to Perihelion, in astronomical units	Distance from the Sun to the Aphelion, in astronomical units	Major Semi-axis of Orbit, in astronomical units	Rotation Period in Years	Number of Ellipses passed between Two Routine Encounters with Earth	Time Interval Elapsing between Two Routine Encounters with Earth, in years	Velocity of Take-off from Pole of Earth, in kilometers per second	Velocity of Take-off from Interplanetary Space Station of Earth, in kilometers per second	Velocity of Take-off in Relation to the Sun outside the Earth's sphere of influence, in kilometers per second	Ratio of the Initial Mass of the Rocket Flying from the Interplanetary Space Station, to its Terminal Mass in case of Gas Escape Velocity at 4 kilometers per second
1	0.260	1	0.630	0.500	2	1	15,458	7,529	10,646	6,568
2	0.352	1	0.676	0.556	9	5	13,944	6,000	8,296	4,482
3	0.377	1	0.689	0.571	7	4	13,617	5,670	7,735	4,727
4	0.423	1	0.711	0.600	5	3	13,119	5,166	6,820	3,638
5	0.462	1	0.731	0.625	8	5	12,761	4,803	6,102	3,323

TABLE IX. (*Contd.*)

6	0.526	1	0.763	0.667	3	2	12,291	4,327	5,047	2,950
7	0.598	1	0.799	0.714	7	5	11,904	3,934	4,013	2,674
8	0.651	1	0.825	0.750	4	3	11,693	3,719	3,333	2,534
9	0.724	1	0.862	0.800	5	4	11,481	3,506	2,492	2,402
10	0.711	1	0.866	0.833	6	5	11,383	3,405	1,990	2,343
11	1	1.321	1.160	1.250	4	5	11,383	3,405	1,991	2,343
12	1	1.423	1.211	1.333	3	4	11,481	3,506	2,493	2,402
13	1	1.621	1.310	1.500	2	3	11,694	3,721	3,338	2,535
14	1	1.811	1.406	1.667	3	5	11,908	3,938	4,024	2,676
15	1	2.175	1.587	2.000	1	2	12,311	4,338	5,076	2,958
16	1	2.684	1.842	2.500	2	5	12,791	4,834	6,165	3,348
17	1	3.160	2.080	3.000	1	3	13,174	5,299	6,923	3,755
18	1	4.040	2.520	4.000	1	4	13,723	5,778	7,929	4,240
19	1	4.848	2.924	5.000	1	5	14,104	6,162	8,562	4,667

the launching of artificial satellites become more graphic from the following examples.

Let us assume, for example, that Canada decided to launch an artificial satellite from Grant Land. Does she have the right to do so? Positively yes, if we proceed from air law. But Grant Land is situated above the 80th parallel, and regardless how an artificial satellite be launched from this territory, it will intersect the Equator without fail and also fly around the southern hemisphere right to the Antarctic. In this case, any of the countries over which the artificial satellite will fly can consider this fact a "violation of its sovereignty". But can a given country prohibit Canada from launching an artificial satellite from her own territory?

Such countries as Indonesia, Brazil, Colombia and others crossed by the Equator, might construct satellites which hover "motionless" over their territories (the artificial satellite space station).

According to existing international legal standards nobody can prohibit, let us say, Ecuador to equip such a satellite slightly north of Quito (capital of Ecuador) above the Equator. But according to the same international law, such countries as Mexico, Bolivia, Brazil and some others might protest: the whole territory of their countries would be in the field of vision of an observer who is on board such a satellite, which will then fly at an altitude of scores of thousands of kilometers.

Jurists thus encounter so far insoluble problems. It is therefore not surprising that at present jurists differ sharply in opinion as to the lawful standards which must regulate questions of the use of world space. Some foreign specialists advance the thesis of "freedom of space" (Meyer, Federal German Republic). Hurford (Great Britain) holds the opinion that from the viewpoint of law outer space should be compared with the open sea in which ships can sail under various flags. Others speak out for restricting the freedom of movement of cosmic vehicles (Cooper, Argentina). Crocco holds the opinion that the sovereignty of the state in respect to air space must be limited to a definite altitude above which space will be free for space navigation, just as the oceans are free for transcontinental communication (but the artificial satellites will themselves belong to a state, a group of states or, finally, to a private international enterprise, which created them).

The main difficulty which jurists confront is this question: To what altitude above the surface of territories subject to them have states the right to maintain their supremacy? Although visible demarcation boundaries cannot be established in space, legislative acts which regulate the upper frontiers of states can proceed from properties of the Earth's field of gravitation or upper layers of the atmosphere, which are subject to exact definition. Such criteria are advanced, for example, as the blue light of the sky which is still observable from the travel altitude of an artificial satellite (Crel*, France) or the "maximal distance from which a body still falls on the territory of a given state" (Cooper), which the author himself does not define with sufficient clarity.

Astronautics will, of course, develop despite all these jurisdictional difficulties. If appropriate measures are not taken, however, space travel development along with countless benefits can bring immense destruction.

The testing of orbital rockets and also other measures in the field of astronautics must, in the opinion of certain jurists, be subjected to international control. It is also necessary to establish a system of artificial satellite travel.

In the opinion of A. W. B. Hester, member of the British astronomical society, the United Nations Organization should engage in working out international standards in the field of astronautics and the testing of ultra-long-range rockets, because negotiations through usual diplomatic channels can hardly lead to positive results. Hester presents the following proposals†: (a) establishment of a United

* The French authority, Crel, whom Sternfeld quotes, could not be positively identified; hence, the name may be spelled differently than it appears here. The same applies to the Argentinian, Cooper. There is a John Cobb Cooper who writes on space law, but it could not be determined whether he is the man to whom Sternfeld refers.

† Translator's Note: The proposals of A. W. B. Hester are presented in the text as author Sternfeld described them, retranslated into English from the Russian. Hester's list of what he termed thorny problems is herewith presented as it originally appeared in the *Journal of the British Interplanetary Society:*

(a) Constitution of the U.N.O. commission for the co-ordination of research and development of astronautics. —

(b) Method of pooling information and the interchange of scientists and technicians.

(c) Freedom of international development and limitation of private ventures.

(d) Control and co-ordination of operations. Under this consideration would be taken schemes for notification of countries to be traversed. Details would include trajectory, velocity and duration of outward and homing flights.

(e) Methods for the reduction of operational hazards. Necessary restrictions

Nations commission for coordination of scientific investigations in the field of astronautics; (b) exchange of information, scientists and technicians; (c) freedom of international measures and limitation of private initiative in the field of astronautics; (d) control and coordination of tests, including a program of selected routes, speeds, times, coordinates of take-off and descent, etc.; (e) methods of limiting the number of accidents connected with take-off, flight and descent of vehicles (so, for example, it is possible to restrict the route of a test flight from one pole to the other to water space and definite territories, to employ preventive appliances against radioactive contaminations, to establish starting platforms in localities with glacial cover etc.); (f) prohibition of the supply of flying machines of any kind with combat loads, the prohibition or at least limitation of the use of astronautical vehicles for war purposes; (g) adoption of sanctions against states which violate the agreement on peaceful use of astronautics.

It is perfectly clear that all the problems posed by the space age, despite their complexity, can be solved if the negotiating sides have a goodwill approach, and on the condition that artificial satellites will be used only for peaceful scientific purposes.

---

and precautions would cover limitations within terrestrial atmosphere, pole to pole sea-traversing trajectories from ice-cap fringe bases, barometric control for systems and introduction of safeguards against radio-active contamination.

(f) The prohibition of war-head explosives and destructive material other than propulsive fuels.

(g) Agreement on joint action by U.N.O. against a defaulting nation. Such defaults might be:

(1) Mis-application of astronautics for military purposes.

(2) Refusal to disclose information within the terms of the internationally agreed policy.

(3) Deliberate violation of a foreign territory by disregard of internationally agreed safeguards and procedure.

Conclusion

IN THE previous chapters, we have made an attempt to give the reader a glimpse into the future of astronautics.

The first artificial satellites of the Earth (Sputnik I and II) were created by the efforts of Soviet scientists and technicians. On the basis of this experience, a series of satellites will be launched somewhat larger in size and equipped with increasingly complex and diversified instruments. It will then be necessary to test whether flight in a satellite vehicle is harmless for the living organism. (Such testing began when the second Soviet satellite was orbited with a dog on board.) Space travel will enter the final stage when artificial satellites are built large enough to accommodate not only instruments but people as well.

The first artificial satellites will fly around the Earth in elliptical orbits more or less close to the surface of our planet. In future higher and higher speeds will be communicated to satellites space vehicles and they will then complete flights in more and more elongated orbits.

Great difficulties will at first confront man is raising the "ceiling" of the artificial satellite, but in proportion to the increase in rocket power this problem will be solved with increasing ease. In fact, an increase of initial speed at the surface of the earth, for example, from 7.9 to 10 kilometers per second will raise the satellite flight "ceiling" by three equatorial radii, while a further increase of this speed by one kilometer a second will result in raising the "ceiling" by 25 radii of the Earth. Thus, at a speed of 11 Kilometers a second the rocket will already fly half the distance between the Earth and the Moon. Flight around the Moon and nearest planets will then follow.

To reach the Moon and all the planets of our solar system the rocket's speed must vary between 11.1 and 16.7 kilometers per second.

The construction of an interplanetary transfer station will greatly facilitate space travel, making it possible for the space ship to attain its necessary speed in two stages. The ship will accelerate to a circular velocity of 7.9 kilometers per second in the initial take-off from Earth, and gain an additional 3 to 4 kilometers per second at the take-off from this station.

Thermochemical-fuel rockets will probably be the first to venture into space but there is no doubt that they will be followed by atomic space ships which will be far superior to them. Atomic energy opens up great possibilities for astronautics.

The atomic rocket will make it possible to fly to the Moon and other planets without stopping at an interplanetary station for refuelling. By using rocket-braking the atomic space ship will be able to land on planets or their satellites devoid of an atmosphere and will also be able to return to the Earth from any planet of the solar system. And last but not least, it will be able to take off without waiting for the most advantageous planetary configuration.

After gaining speed the space ship will fly under its own momentum to save fuel, and for the same reason it will not follow a straight course in space. Its trajectory will be an ellipse, and later a parabola or hyperbola.

Unmanned radio-controlled rockets will be launched to explore space before man is prepared to follow them. They will collect data required for the building of a space ship and test the conditions of space-flight on animals.

The building of an artificial satellite of the Earth will be the first stage of interplanetary travel, to be followed by a trip to the Moon and other planets.

A round-the-Earth flight will take not more than one and a half hours. A flight around the Moon and back to the Earth will take ten days, whereas a journey along an elliptical trajectory crossing the orbits of Venus and Mars, including a homeward journey, will take not less than one year. Expeditions to the more remote worlds will take several years.

Modern radio engineering will provide communication facilities by using directed radio waves. The positions of rockets flying into space will be easily determined at any time because rockets are subject to the same laws as celestial bodies.

As far as can be visualized, there is no obstacle to interplanetary travel from the physiological point of view. In all probability man will be able to bear the strain amounting to four or five times the weight of his body during the few minutes the motors are running, which means that the space ship will be able to gain cosmic speed with its rockets working under the most economical conditions.

As far as weightlessness is concerned, it is not yet definitely established that it will produce no harmful effects on the human organism during a more or less protracted period of time. However if weightlessness turns out to be harmful it can easily be counteracted—it is technically possible to create artificial gravitation by rotating the space ship.

The temperature of the air inside the space ship will be controlled by the ship's plating, which will absorb solar energy more or less intensively.

At the present stage of technical progress there is no special difficulty about creating a microatmosphere in the space ship of a composition and humidity to suit man, or supplying the astronauts with food and protect them against ultra-violet rays from the Sun. The effects of cosmic radiation on the human body are being studied. The fact that meteors and asteroids will be very dangerous to space ships has to be borne in mind.

The latest scientific achievements make it possible to draw the conclusion that interplanetary travel can be accomplished during the present century. The great dream which not so long ago was considered to be fantastic is now coming true.

Interplanetary travel will shed light on the hitherto unanswered question of whether there is life, and if so, how advanced it is, on the other planets of our solar system.

Apart from its purely scientific interest space travel will probably be of practical value, although at this stage it is difficult to specify in what way. By way of illustration we can point to the fact that the planets and their satellites are an inexhaustible source of mineral wealth which must be studied and utilized for the well-being of mankind.

Soviet people will build interplanetary stations and space ships in order to uncover the secrets of the universe and extend the domain in which human reason reigns over the elements.

part two

The Sputniks

Space Travel Initiated

by Academician V. AMBARTSUMYAN
Byurukan Astrophysical Observatory

MANY TIMES in the history of science major discoveries or inventions, the fruit of the paintstaking labor and profound study of a whole team of scientists, have revolutionized further scientific progress. Still we find it hard to give any past example, when one experiment—true, the result of a major effort by a large group of scientists, engineers, technicians and industrial workers and the reflection of tremendous scientific and technical progress—would usher in a new epoch in the history not only of science, but of all of human culture and engineering. But that was what happened on October 4, 1957, when for the first time in history a *new astronomical body* traveling around the Earth, its first artificial satellite, was introduced, to thereby initiate astronautics.

The importance contained in the launching of the man-made Earth satellite for technology and engineering is immeasurable. This development spells a transition to cosmic velocities which can overcome gravitation. Thus, free flight without fuel in airless space becomes feasible.

The artificial satellites are particularly important for further scientific advancement.

The satellite moves within the field of terrestrial gravitation. In its turn this field is determined by the distribution of masses inside the earth and in the earth's crust. By studying the satellite's motion we can make our knowledge about the field of terrestrial gravitation far more accurate and, hence, draw interesting conclusions about the Earth's structural composition. True, the Earth is girdled by a natural Moon, but a study of its motion furnishes information only about sections of the field of terrestrial gravitation that are compartively far removed from the earth, since a distance of some 380,000 kilometers (235,980 miles) lies between the Earth and the Moon. At such distances the field of terrestrial gravitation is far less consequent upon the distribution of masses inside the Earth itself. Therefore, a study of the Moon's motion can yield only niggardly information on this score. On the other hand, artificial satellites circling around 1,000 kilometers (621 miles) away from the Earth give us far more opportunities in this respect.

At several hundred miles up, the atmosphere is extremely rarefied. Nevertheless, the air must give some resistance and influence the satellite's motion. Therefore, by studying this motion we shall add to our store of knowledge about the composition of the top layers of our atmosphere.

The tasks listed above can be achieved even if the satellite has no instruments at all, for in this case we need only locate accurately the changing bearings of the satellite. It goes without saying though, that the satellite's equipment for the transmission of radio signals earthward makes location of the satellite's bearings much easier.

The range of questions that can be tackled by equipping the satellite with the appropriate scientific measuring instruments that would automatically radio measurement results is still wider.

When we study the celestial bodies around us and the cosmic space in which the earth revolves, we astronomers encounter great difficulties due to our observatories and scientific stations being placed at the bottom of the ocean of air that envelops the Earth, an ocean hundreds of miles deep. This ocean of the Earth's atmosphere lets through only isolated, narrow sectors of the spectrum of electromagnetic oscillations emitted by the Sun, the stars and other celestial luminaries. We have therefore always dreamed of an observatory outside the atmosphere from which we would be able to unhamperedly observe ultraviolet, x-ray solar radiation, radio emissions from a long wave of the order of several dozens or hundreds of meters, and also the charged particles the Sun emits, especially at times of great activity. Finally, the cosmic rays born in remote nebulae that throughout the outer space arround us undergo a whole series of transformations upon entering our atmosphere, making it very difficult for us to gauge the character of cosmic rays in their pure state. The instruments on the satellite will enable us to investigate pure cosmic rays.

An investigation of the ultraviolet sector of the solar spectrum and also of the charged particles the Sun emits is of particularly great importance. These radiations powerfully affect the state of the upper layers of the Earth's atmosphere, to wit, the ionosphere. They condition its ionization, which determines the basic properties of the ionosphere linked with the reflection and passage of radio waves, with auroras, etc. However, the intensity of these solar radiations is subject to radical changes. For that reason the state of the ionosphere and its properties also change all the time. To elucidate the natural laws of development governing these changes we need exact and direct data on the above-mentioned solar radiations. This can be derived through extra-atmosphere observation.

On the other hand, the installation of radio transmitters on man-made satellites will allow us also to directly study the ionosphere. This is because reception of the signals these transmitters send by numerous ground-borne receiving stations and determination of the intensity of the signals received will enable us, so to speak, to feel through the ionosphere the medium via which these signals pass.

Finally, so far we could obtain our information about the medium filling interplanetary space, meteors and interplanetary gases only by observation "from afar," *i.e.*, from ground borne observatories. The man-made satellite, however, contacts this medium directly. Therefore, by equipping the satellite with the appropriate instruments we shall be able for the first time to obtain data directly bearing upon the interplanetary medium.

It goes without saying that to fully solve the tasks listed we shall need not one but many satellites with different and even better instruments, as well as an even better method for automatically telemetering scientific information earthward. In its turn the data obtained with the help of the satellites will naturally pose a large range of new problems.

However, it is easy to see that the launching of the first man-made satellite is gladdening millions not only because it brings nearer a solution to the questions listed above. Apart from its superlatively colossal scientific importance, the launching of the artificial satellite is of incalculable value as being the first step into outer space, the debut to space travel, and the first stage toward realization of the dream of man's conquest of the boundless expanses of the universe.

We may predict the construction within the next few years of satellites that will be able to circle around the Earth several thousands of miles away, which will be equipped with instruments for every kind of scientific measurement.

The next step should be to make a rocket to overcome terrestrial gravitation, reach the vicinity of the Moon and fly around it. This rocket would furnish a wealth of information about the nature of lunar terrain and tell us about the structural constitution of the lunar hemisphere we never see. This would be followed up by vehicles for interplanetary travel.

We can, of course, differ on how soon man will be able to embark upon space voyages. But the point is that the development of modern automation and electronic computers has reached the stage where we can, in principle, construct a machine which, in addition to taking measurements, could also rationally, without human guidance, decide what measurements should be taken and the order in which they should be taken, depending on the results of earlier measurements. Radio equipment, moreover, makes it possible to automatically and quickly relay earthward the results of a vast number of observations and measurements. It is also possible in principle to televise what the flying machine sees. Thereby, the opportunity arises for vastly extending the functions of automatic observatories circling in space outside the atmosphere. Without question, further technical progress will enable us to launch manned vehicles into outer space.

Pravda, October 11, 1957

From Man-Made Satellite To Moon-Bound Voyages

by Professor V. DOBRONRAVOV
Doctor of Science in Physics and Mathematics

THE MAKING and launching of the world's first test Earth satellite (Sputnik) is a superlative achievement for Soviet science and engineering. All branches of knowledge and a wide range of different industries were represented in the designing, making and launching of Sputnik and its carrier rocket, and in the computation of their further movement.

We can already formulate a number of principles stemming from an analysis of all the phenomena involved in the launching and movement of the satellite.

We must note that the satellite's behavior has confirmed all the forecasts of its makers with respect to the orbit and the evolution of this orbit, and also with regard to the physical situation of the environment through which the satellite is passing.

Indeed, the satellite began to move along an orbit computed beforehand in line with the laws of rocket dynamics and astronomy. Its braking and dropping to shorter orbits and the consequent decrease in the orbital period of revolution are also conforming to previous expectations. Apparently, the satellite has not yet met any large meteor particle. The density of atmospheric matter and the distribution of micrometeoric dust in the regions of space through which it is passing are evidently close to the data the scientists had even before they launched Sputnik.

The life of the satellite appears to be continuing for a long while, and this again dovetails with the reckonings made beforehand.

As for the specific structural features of the satellite, we should undoubtedly note its large weight, that is, large in relation to its volume. This implies the successful planning of equipment and the large capacity of this equipment despite the comparatively small size. The system of aerials mounted on the satellite in the form of rods is also most efficient. When the carrier rocket was coursing toward the projected orbit, these rods were pressed to the sides of Sputnik. However, when the satellite parted from the carrier rocket, the rods moved out and turned slightly on their swivels to take up a definite and permanent position in relation to the satellite.

We must certainly emphasize the reliability of the radio equipment and the great capacity of the power resources aboard Sputnik. The satellite is still sending out signals regularly on the set main frequencies and consequently ground stations are getting regular radioed results of different observations, as, for instance, of cosmic particles. They come in the form of signals which can be analyzed according to the laws of the theory of information, a division of that new modern and vast science of cybernetics.

Finally, we must certainly draw attention to the fact that the carrier rocket has also become an earth satellite, since at the end of the transitional trajectory it had worked up a velocity sufficient to enable it to follow the satellite along short orbits. This is evidence of Soviet rocket technique's being in possession of engines so powerful that they hold out the prospect of the development of still larger satellites for launchings at still greater distances from the Earth.

Today we can speak with definite conviction about the possibility of making a flying machine to reach the Moon.

Many people would like to know whether the satellite has a rotation of its own. It is quite likely that there is some slight rotation around the center of the mass of the entire system, that is to say, the satellite and its rods, for when the satellite was thrust out of the carrier rocket, there may have been some assymetry in the forces operating upon

Sputnik from the thrusting device. This could have caused an original moment of rotation imparting an original impulse to the satellite itself. In turning on their swivels the rods could have also lent an original rotation to the satellite body.

Any further development and improvement in the making of artificial satellites should apparently follow the policy of increasing their size and weight, for this would permit still greater power resources for radio communication and the operation of scientific instruments.

For more effective observation the satellite should be "oriented" in space in one way or another. Consequently, it is desirable for it to have a device on board that would automatically control its flight.

The next stage apparently will be the making of radio-controlled satellite-rockets possessing engines and definite stocks of fuel. By automatically turning the engines on or off we shall be able to switch the rocket from one orbit to another. Automatic devices to brake the rocket upon its entry into a more dense atmosphere are thought to be expedient, for then the satellite would be able to return to the earth, and the results of scientific observation would be still more valuable.

At a later stage such living creatures as dogs and monkeys could be sent out aboard the satellite to girdle the Earth, enabling us to see how these creatures behave at different stages of the transitional trajectory and also during flight along the satellite's orbit.

As the satellite circles the Earth, all objects inside of it will be subjected to the influence of the centrifugal force of inertia resisting the Earth's gravitational pull and decreasing the latter. A living organism will find itself without weight. It is important to know how it will behave under such weightless conditions in order to be able to plan future space travel for man.

Observation of animals will also help us to find how blood circulation and the alimentary processes react in the living organism when in a weightless state.

Man will board rockets and satellites after rockets are equipped with well-regulated engines, pilot personnel are ready and trained to withstand the weightless state and after special space suits, capable of preserving life in airless space, are made.

Manned rocket satellites will be able to switch from one orbit to another and to link up with each other. The crew will be able to emerge onto the outer surface of the flying machine, thus making possible the establishment of large, earth-girdling "space stations," of which K. E. Tsiolkovsky himself spoke. These flying stations could provide the take-off field for distant space voyages and also be used for the assembling of space ships.

Further, we ought to note that voyages to the Moon, Mars and other planets will be mainly "ballistic," i.e., the space ships will fly with their engines cut off, acting under the pull of one or another attracting center of gravitation. Engines will be turned on only for high-speed maneuvering, in switching from one course to another and for take-offs and landings.

To enable a body to come within reach of the Moon's gravitational pull, it must have a speed of roughly seven miles per second within the neighborhood of the Earth. Then it will move along a greatly extended ellipse and reach a zone where the Moon's pull will act on it more strongly. By maneuvering, i.e., by turning on the engines at different times, the space ship will be able to take a circular course and fly around the Moon, land on it, ascend from it or again revert to the elliptical orbit for the return trip to the Earth.

Some of the routes for a Moon-bound flight of different durations, from one to five days (one way only), have already been computed.

Of course, at first, as a preliminary stage, distant interplanetary space will be reconnoitered by radio-controlled flying machines. It is thought that maiden voyages to the Moon in particular will be made by unmanned rockets, outfitted with various automatically-operated instruments for the purpose of photographing the side of the Moon we never see and for televising earthward pictures thus made. The making of different cybernetic mechanisms or "robots" capable of automatically alighting from the ship onto the Moon's surface, conducting some observations and relaying them to earth stations is also quite feasible.

The launching of the artificial earth satellite is man's debut into interplanetary space. The not-so-distant future will show how man is going to conquer space.

Promyshlerno-Ekonomicheskaya Gazeta, October 20, 1957

Earth Satellites And Geophysical Problems

Interview with ALEXANDER OBUKHOV
*Corresponding Member of the USSR Academy of Sciences
and Director of the Institute of Atmospheric Physics*

SOVIET SCIENCE and engineering have achieved another tremendous victory by launching a second sputnik into the cosmos. Soviet scientists are expanding research of cosmic space and the upper strata of the atmosphere. The unexplored processes of natural phenomena taking place in the cosmos will become now more accessible to man. At the present time Soviet scientists are able to make observations of the globe from a cosmic scientific observatory, an unprecedented event in history.

What geophysical problems can be solved, and are already being solved, with the aid of the sputniks?

Observations of the trajectory of the sputniks' flights make it possible, first of all, to considerably enhance our data about the density of the Earth's atmosphere at great altitudes. The initial studies of the radio signals coming from the sputniks already provide us with extraordinary valuable information on the electric properties of the upper strata of the atmosphere, i.e., the ionosphere.

The direct study of the nature of the upper layers of the atmosphere with the aid of special apparatus with which the sputniks have been equipped is of great scientific and practical interest. Suffice it to recall that the broadcasting of radio signals at great distances completely depends on the processes taking place at altitudes exceeding the 60-kilometer mark. In order to be able to forecast the distribution of radiowaves, it is necessary to known the physical factors that determine the degree of the conductivity (ionization) of the atmosphere at these altitudes. The observations conducted with the aid of the spuniks make it possible not only to study the properties of the upper atmosphere, but also discover the reasons for the changes in the ionosphere's characteristics.

The modern conception is that the causes for the aurora borealis, the rapid changes in the Earth's magnetic field and in the conductivity of the ionosphere are minute particles—corpuscles radiated by the Sun. They interact with the magnetic field of our planet. Now the nature of these particles, as also the energy and some other characteristics of their motion, can be determined.

The sputniks open up new vistas in the study not only of the properties of the upper atmosphere, but also of its lower strata, where weather is formed. Even such a comparatively simple, it would seem, problem as the distribution of clouds on the territory of the globe and, connected with it, the distribution of air currents has been little studied to date, for there arise certain difficulties in its solution only by "ground means." It should be borne in mind that a large part of the Earth's surface, such as the oceans and the polar regions, are not very accessible to stationary observations.

The sputniks follow a very intricate trajectory, which almost covers the entire territory of the Earth. Thus it will become possible to receive detailed information on the distribution of clouds, and consequently also air currents, over the entire surface of our planet. This is of major importance for the study of the general circulation of the Earth's atmosphere and the creation of physically grounded methods of long-range weather forecasting.

With the aid of the sputniks the cloud systems throughout the globe may be photographed, their movements registered, etc. All this will give new, valuable data for weather forecasting. The sputniks will also help investigate many optical properties of the atmosphere: light diffusion in it, the absorption and refraction of solar radiation, etc.

It is very important that the sputniks offer an opportunity of conducting a systematic study of the geophysical phenomena in time and space.

Charts of the income and expenditure of solar energy in various parts of the globe drawn up with the aid of the sputniks will also be of great scientific interest. This is but a small part of the huge complex of scientific investigations which will be solved by studying the observations made with the aid of the sputniks. We could, for example, mention the study of meteoric dust filling interplanetary space, the anomaly of the earth's gravitational field and other interesting scientific problems.

Sputnik II

Article in Pravda, November 13, 1957

As HAS already been reported in the newspapers, in line with the research program being carried on under the IGY, on November 3 the Soviet Union launched its second sputnik. This was a new signal victory for Soviet science. As the result of the hard and successful work of a large team of scientists, engineers, technicians and factory workers, a sputnik with a payload of 508.3 kilograms (1120.29 pounds), six times heavier than Sputnik I was made and aligned on orbit. This orbit is, moreover, much further away from the Earth than that of Sputnik I.

Sputnik II is equipped with a diverse package of scientific apparatus enabling a wide program of investigation to be carried out. This includes instruments for the investigation of cosmic radiation and the ultraviolet and X-ray regions of the solar spectrum, a hermetically sealed chamber with a test animal inside (a dog), radio telemetering equipment for relaying the results of measurements to ground stations, radio transmitters and also the required number of electric batteries.

The Sputnik's Orbit and Its Evolution

Sputnik II was aligned on its orbit by means of a multiple-stage rocket. In the process of alignment the rocket rose to a height of several hundred kilometers away from the ground, and then in its last stage turned to move parallel to the ground at a speed of more than 8,000 meters (26,240 feet) a second, becoming a sputnik. At the moment of alignment on orbit the stocks of fuel in the rocket's tanks had been expended and the engine had burned out. The sputnik continued to move due to the kinetic energy acquired from the take-off.

The speed imparted to the final stage of the rocket was greater than the velocity required to have the sputnik travel along a circular orbit at a constant altitude of that of the point of alignment on orbit. For that reason the sputnik follows not a circular orbit but an elliptical orbit of which the apogee is about 1,700 kilometers (1,055 miles), almost double the apogee of Sputnik I. Since the size of the major semi-axis of the orbit of Sputnik II is greater than that of Sputnik I, the period of its revolution around the Earth is also greater, being 103.7 minutes at the beginning of its flight.

Owing to the increased period of revolution, Sputnik II makes about 14 full revolutions around the Earth in 24 hours, whereas Sputnik I made about 15 revolutions at the very beginning. The shift of each turn of the spiral along the line of longitude due to the Earth's sidereal rotation is roughly one-fifteenth larger for Sputnik II than for Sputnik I. The distance apart between two turns of the Earth spiral has increased accordingly by the same value.

The resistance of the Earth's atmosphere has had a braking effect on the sputnik, and its orbit is thus changing in size and shape. Since the atmosphere is exceedingly rarefied at great altitudes, the braking effects acting on the sputnik are negligible. Therefore the changes in the orbital parameters are very slow. Since atmospheric density rapidly diminishes as height increases, the main braking effect is in the region of the perigee, that is, at the point closest to the Earth. At the apogee, the furthermost point away, the sputnik moves at such a great altitude that it is already in outer space, outside the Earth's atmosphere which is theoretically conjectured to reach up to a height of the order of 1,000 kilometers (621 miles) away from the Earth.

The slowing down of the sputnik depends not only on atmospheric density but also on the sputnik's shape and the ratio of its weight to the sectional area (on the so-called transverse load). When the transverse load is greater, the loss in speed is less.

Two sputniks, established at the outset on one and the same orbit, but possessing a different braking value, will, within a certain period of time, move differently, as the orbits of their flight will change with different velocities. Here the shrinking of the orbit will be chiefly due to the lowering of the apogee.

At the outset Sputnik I and its carrier rocket traveled approximately along one and the same orbit. The difference between their periods of revolution was negligible, the period being about 96.2 minutes. But now, due to the slowing down of Sputnik I being less than that of the carrier rocket, their orbits essentially differ. The apogee of the carrier rocket is more than 100 kilometers (62 miles) lower than that of the sputnik. According to data for November 10, 1957, the carrier rocket's period of revolution was already roughly 74 seconds less than that of Sputnik I.

The braking value for both the carrier rocket and the sputnik changes with time, due to the changes in the orbital parameters. As the orbit goes down, the braking increases in progression. This is lucidly corroborated by the results of observation. When the orbit falls to the height of the order of 100 kilometers, the braking will be so great that the sputnik and the carrier rocket will be intensively heated up and then continue to drop rapidly and burn up.

The sputnik's life-span depends on the value of the atmospheric braking. It is quite clear that the greater the period of revolution and the less the braking, the longer the sputnik's life. Calculations based on data yielded by tracking Sputnik I and its carrier rocket warrant our presuming that the sputnik's life will be about three months, counting the day of its launching. This means that Sputnik I will evidently travel along its orbit until the end of this year. The life of the carrier rocket will be less than that of Sputnik I, and therefore we ought to expect the carrier rocket to burn up before the sputnik. The great period of revolution of Sputnik II and the little importance of the braking value, which is less than for Sputnik I, warrants the assertion that the time Sputnik II will travel along its orbit will be appreciably longer than that of Sputnik I.

Present systematization of the results of trajectory measurements will enable us to determine fully the entire process of the evolution of the orbital parameters of both sputniks and obtain important information about density distribution in the top layers of the atmosphere. Later, we shall be able to reliably forecast how long sputniks will live.

Observation of the Sputniks

Sixty-six special optical observatories, every astronomical observatory of the Soviet Union, and close to thirty foreign observatories are tracking the first two sputniks and the carrier rocket of the first sputnik. At present a whole network of optical observatory stations is being set up in the People's Democracies. The number of foreign astronomical observatories taking part in the systematical observation of the sputniks is increasing with each passing day. The brightness of the carrier rocket and the second sputnik have made it possible to include aerological stations of the Hydrometerological Service which have piloted theodolite balloons in the system of visual observation.

As a result of optical observation, it has been determined that the carrier rocket changes its brilliance. This is brought about by changes in its orientation in outer space. The shortest registered visual period of the shine is approximately 20 seconds.

Apart from visual observations, photographical observations of the carrier rocket and the second sputnik are also taking place. Photographs taken at the Pulkovo Observatory, the observatory of the Astrophysical Institute of the Academy of Sciences of the Kazakh Republic, the observatory of the Kharkov State University and in other astronomical observatories of the Soviet Union, as well as photographs taken at the Crimson Mountain Observatory (People's Republic of China), the Edinburgh Observatory (Great Britain), the Dunsink Observatory (Ireland), the Potsdam Observatory (German Democratic Republic) and others have made it possible to define more exactly the orbits of the sputniks and the carrier rocket.

Radio observations of the sputniks have provided a mass of important data. These observations were done at stations located at different latitudes and longitudes, by radio direction finders, DOSAAF clubs (Voluntary Societies to Assist in Defense), by several technical institutes and thousands of radio hams. The material amassed is so great that only preliminary analysis has been carried out at this time.

Of great importance are the measurements of the intensity of the field of radio signals received from the sputniks. These measurements were made by continuous automatic writing machines and by individual measurements taken at definite intervals. The results of the measurements of the intensity of the field of radio signals have made it possible to evaluate the absorption of radio waves in the ionosphere, including those areas which lie above the maximum ionization of the main ionospheric layer F_2 and therefore are inaccessible to ordinary measurements taken from the earth's surface. These measurements permit us to judge the possible means of transmitting radio waves in the ionosphere.

The results of radio signals received from the sputniks and the measurements of their levels has determined that these signals on the 15-meter band were received at distances far exceeding those determined by direct vision, for they were as great as 10, 12, and even 15,000 kilometers, and in some cases even greater.

The fact that the sputnik, moving on an elliptical orbit, occupies various positions in relation to the basic maximum of electronic concentration in the Earth's atmosphere is of great interest. In analyzing the materials gained through radio observation, it is of importance to ascertain whether, at a given time, the sputnik was above or below the true height of the maximum of electronic concentration of layer F_2, information that is obtained on the basis of high-frequency characteristics of the ionosphere recorded by ionospheric stations. If, in the Southern Hemisphere, the sputnik moves above the ionosphere, then, in the Northern Hemisphere, at times it is above the maximum ionization of this layer, at other times it is lower, and at still other times it is close to this maximum. Such conditions create a great variety in the means of propagating short waves over great distances. One such alternation is the reflection from the earth's surface of radio waves coming from above that have passed through the layer of the ionosphere which are then followed by a single reflection from the ionosphere in areas where the threshold frequencies are of sufficient value. In other instances, radio waves coming from above at a certain angle to the ionosphere are considerably refracted and therefore penetrate this area which lies beyond the limits of straight geometrical vision.

When the sputnik is close to the area of maximum ionization of the atmosphere, conditions are extremely favorable for propagating radio waves through ionospheric radio-wave transmitters. In several cases, as practice has shown, the radio waves pass into the receiver after having circled the earth along a greater orbit instead of following the shortest possible course. In several cases the radio signals produced a worldwide echo. In other cases the measurements of the intensity of the field were greater than they had been thought to be, according to the law of inverse proportionality of the first stage of the distance which also indicates the existence of wave-transmitting channels in the ionosphere.

Interesting results have been obtained on the basis of Doppler's Effect by tape-recording the changes in the tone of beats between the frequency of radio waves sent out by the sputnik and the frequency of vacillations of the local heterodyne. Many such tape recordings have been made and their results are now being analyzed.

Undoubtedly, the final analysis of data based on radio observations of the sputniks will bring about important discoveries of the peculiarities in the ionization of the upper areas of the ionosphere and also of the absorption and nature of the radio waves propagated therein.

Structure of Sputnik II

As has been indicated above, Soviet Sputnik II, unlike Sputnik I, is the last stage of a rocket housing all the scientific and measuring equipment. This way of housing equipment appreciably simplified the task of ascertaining the sputnik's bearings by means of optical observation, since, as the experiment with Sputnik I has shown, the tracking of the carrier rocket proved to be far simpler than the tracking of the sputnik itself. The brilliancy of the carrier rocket is greater than that of Sputnik I by several stellar magnitudes. The total weight of the equipment, the test animal and the electric batteries in Sputnik II is 508.3 kilograms (1118.26 pounds).

In the nose of the final stage of the rocket, fixed to a special frame, are instruments for studying solar radiation in the ultraviolet and X-ray regions of the spectrum, a spherical container with radio transmitters and other instruments and a hermetically sealed chamber with a test animal (a dog) inside. The apparatus for cosmic ray investigation is fixed to the rocket's body. The devices and containers installed in the rocket are shielded from the aerodynamic and heat effects of the rocket's passage through the dense layers of the atmosphere by a special protective cone. After the last stage of the rocket was established on its orbit, the protective cone was discarded.

The radio transmitters, placed in the spherical container, functioned on frequencies of 40.002 and 20.005 megacycles per second. The electric batteries, the heat regulation arrangement and also sensitive elements for recording temperature fluctuations and other parameters were also placed in this container. In structure the spherical container resembles the first sputnik.

The signals emitted by the radio transmitter working on 20.005 megacycles (15 meters) were of a telegraphic order averaging around 0.3 seconds in length, with a pause of the same time in between. Due to changes in some of the parameters inside the spherical container (temperature and pressure), the length of the signal and the pause also changed within definite limits.

The transmitter working on 40.002 megacycles (7.5 meters) functioned under a continual emission regime. The conformity of the two transmitters to the indicated frequencies ensured investigation of the propagation of the radio waves emitted from the sputnik and the measurement of its orbital parameters. Moreover, reception of signals from the sputnik in all ionospheric conditions was guaranteed. The choice of the wave lengths and also the ample power of the transmitters enabled the broadest community of amateur radio operators to track the sputnik along the stations specially assigned to this task.

The hermetically sealed chamber, in which the test animal (a dog) was housed, was cylindrical in shape. To provide conditions for maintaining normal life, it had a stock of food and an air-conditioning arrangement consisting of a regeneration installation and a heat regulation system. Along with this, the chamber also had inside equipment for recording the pulse beat, the breathing and blood pressure, instruments for taking electrocardiograms and also sensitive elements for measuring a number of parameters characterizing the conditions inside the chamber (temperature, pressure).

Both the chamber for the test animal and the spherical container were made of aluminum alloys. They had a polished surface, specially treated to impart the required coefficients of emission and absorption of solar radiance. The heat-regulating arrangement inside the spherical container and the animal's chamber maintained a set temperature, deflecting the heat into the hull by compulsory gas circulation.

Apart from the indicated equipment, the body of the last stage of the rocket had on it radiotelemetering measuring equipment, instruments for temperature measurement and electric batteries to feed the scientific and measuring instruments. The temperature on the outer surface and inside the animal's chamber and also of the individual devices and elements was ascertained by means of temperature regulators fixed onto them. The radio-telemetering equipment relayed to ground stations all recordings of all the measurements taken on the sputnik. The broadcasting of this data was done periodically according to a special programmed arrangement.

The research program involved in the taking of measurements by Sputnik II was to cover a period of seven days. To date it has been carried out. The sputnik's radio transmitters and also the radiotelemetering equipment have stopped working. The further tracking of the motion of Sputnik II to study the characteristic features of the top layers of the atmosphere and forecast its flight is being done optically and by radar.

Scientific Measurement by Sputnik

The sputnik enabled scientists to conduct for the first time several experiments in the top layers of the atmosphere, experiments previously unthinkable.

The Sun's Short-Wave Radiation

Investigation of the short wave ultraviolet radiation of the sun is of paramount scientific and practical interest for physics, astrophysics and geophysics. As investigation carried on in recent years has shown, apart from the visible light, the Sun emits a radiation encompassing a wide range of wave lengths, from X-rays with a wave length of several one hundred-millionths of a centimeter to radio waves several meters in length.

The emission of the short-wave region of the solar spectrum (remote ultraviolet and X-ray radiation) and also of radio waves is linked with the physical processes taking place in the exterior layers of the Sun's atmosphere (the chromosphere and the corona, regions about which we have little knowledge) and has the most serious effects on the Earth's atmosphere. Investigation of the Sun's chromosphere is concentrated mainly on the spectral line of hydrogen, at 1,215 angstroms (an angstrom is one hundred-millionth of a centimeter), located in the remote ultraviolet region of the spectrum. The study of the corona is limited to the region of soft X-rays (from 3 to 100 angstroms). The corona, which is comprised of extremely rarefied matter, has a temperature close to 1,000,000 degrees Centigrade, but, furthermore, apparently the corona has regions with a still higher temperature. To this day the nature of the corona is still much of an enigma.

The aggregate energy of the Sun's short-wave radiation is comparatively small, being tens of thousands of times less than the energy the Sun emits as visible light, but it is this radiation that very greatly affects the Earth's atmosphere. The reason is that short-wave radiation is exceedingly active and can ionize air molecules, thus engendering the ionosphere, which is the strongly ionized upper layers of the atmosphere. According to existing theories, the nether layer of the ionosphere which stretches at a height of between 70 to 90 kilometers* (the D layer) is formed through the ionization of air molecules by the emission of the spectral line of hydrogen from the chromosphere, while the next layer, at the height of between 90 to 100 kilometers** (the E layer), owes its ionization to the corona's X-ray emission.

The condition of the exterior layers of the Sun and of the ionosphere is not constant, it changes continually. It has been found that there is a close link between solar activity, the appearance of so-called chromospheric eruptions, and ionospheric absorption of radio waves, resulting in the interruption of radio communication. This compels the assumption that there is a direct bond between the variations in the intensity of the Sun's X-ray radiation and ionospheric processes.

The Earth's atmosphere fully absorbs the Sun's ultraviolet radiation, attracting only a region of proximal ultraviolet radiation contiguous to the violet in the visible spectrum. These absorbing effects of the terrestrial atmosphere shield living organisms from the Sun's deadly shortwave radiation. But at the same time it makes investigation of this radiaion from the ground impossible. The absorption by the molecules of air is so great that to observe this short-wave radiation we must be altogether outside the Earth's atmosphere and

* Approximately 44 to 56 miles.
** Approximately 56 to 62 miles.

place our instruments on a sputnik. Though the use of high-altitude rockets has also yielded valuable results, only the sputniks enable us to carry on systematic measurement over long periods of time necessary to study the variations in the intensity of short-wave ultraviolet radiation.

Three special photoelectronic multipliers, set at an angle of 120 degrees to one another, serve as receptors of radiation. Each photomultiplier is consecutively covered with several filters of thin metallic and organic film and also special optical material, making it possible to single out the different bands in the X-ray region of the solar spectrum and the hydrogen line in the remote ultraviolet region. The electric signals given out by the photo-multiplier trained on the Sun were amplified by radio arrangements and telemetered to ground stations.

Owing to the continual change in the Sputnik's bearings with respect to the Sun and also to the fact that part of its orbit was not lit by the Sun for a certain period of time, the electric circuits in the apparatus were switched on, in order to feed the batteries, only when the sun came within the range of vision of any of the three light receptors. This was done by means of photo-resistors lit by the sun at the same time as the photo-multipliers, and by an automatically-working arrangement.

Along with sputnik observation of solar radiation, this was also done by the entire network of ground "Sun service" stations incorporated in the IGY program. These observations were carried on by astrophysical observatories and by stations studying the ionosphere and receiving the sun's radio waves. Systematization of all these observation materials will enable us to draw the first conclusions as to the connection between the sun's ultraviolet and X-ray radiation and the processes taking place in the sun's chromosphere and corona, and likewise the state of the earth's ionosphere. This data will pave the way for subsequent systematic observation.

Cosmic Radiation Study

The atomic nuclei of different elements are accelerated and acquire very great energies in outer space. The cosmic radiation that thus originates affords the opportunity of investigating outer space at great distances away from the earth and even from the solar system itself. As they travel toward the Earth from their birthplace, the cosmic rays undergo the effects of the surroundings through which they pass. Due to a whole number of processes, the constitutional structure and intensity of this radiation change. In particular, the number of cosmic ray particles will increase if there be intensive explosions on the sun and if the conditions are created for accelerating atom nuclei to great energies. Thus, there arises an additional stream of cosmic rays, sent out by the Sun.

The Sun is also a source of corpuscular radiation. In the streams of corpuscular rays there are intensive magnetic and electric fields which affect cosmic radiation. Cosmic rays assist us in studying these streams at great distances from the Earth.

Cosmic ray particles are strongly deflected in their passage through the Earth's magnetic field. Only particles with very great energies will reach unhampered any spot on our planet. The less the energy of the particles, the smaller the places on Earth accessible to them. Particles with small energies reach only the polar regions. Thus, the Earth is surrounded in its way by an energy barrier which furthermore decreases in height from an equatorial peak as geomagnetic latitudes increase. Equatorial regions are reached only by cosmic protons with energies of more than 14,000 million electric volts. The southern parts of the Soviet Union are accessible only to particles possessing energies of more than 7,000 Mev. Finally, the neighborhood of Moscow is reached by all particles with energies of more than 1,500 Mev. The measurement of cosmic rays at different latitudes enables us to ascertain how many particles there are in the cosmic radiation and what energies they have. The dependence of the number of particles in cosmic radiation on the latitude, the so-called latitude effect, determines the distribution of the particles according to their energy, i.e., the energy spectrum of cosmic radiation.

Due to a number of processes taking place in outer space and involving cosmic rays, their number and constitution change. In some cases, as, for instance, when particles appear on the sun, there are reasons to expect that only the number of particles with small energies increases, whereas the number of particles with high energies remains without change. In contradistinction, changes in the earth's magnetic field and the effect of the sun's corpuscular radiation on cosmic rays change not only the number of particles with small energies but also the number of particles with high energies.

To elucidate the nature of the changes cosmic rays undergo, we must not only establish the fact of the increase or decrease in the intensity of cosmic radiation, but also determine how the number of particles with different energies will change. Moving at a speed of eight kilometers (approximately five miles) per second, the sputnik crosses from one latitude to another in a very short space of time. Thus, by sputnik measurements of cosmic radiation we can ascertain its latitude effect and thereby the distribution of particles of this radiation according to their energies. The fact that there are a great number of these measurements is particularly appreciated. Therefore, the sputnik will help us trace the changes not only in the intensity of cosmic radiation but also in its composition.

The particles comprising cosmic radiation are recorded aboard the sputnik by means of counters of charged particles. When the counter traps an electrically-charged particle, a spark arises sending an impulse to a radio arrangement of semiconductor triodes which count the number of cosmic ray particles and send out a signal when a definite number of particles have been counted. After the signal is broadcast that a definite number of particles have been counted, the cosmic ray particles are again recorded, with a new signal sent off as soon as the same number of particles have been counted. By dividing the number of recorded particles by the time in which they were counted, we shall obtain the number of particles the counter traps each second, or in other words, we shall know the intensity of cosmic radiation.

The second sputnik had two identical devices for recording charged particles. The axes of the counters in both devices were perpendicular to each other.

Preliminary systematization of sputnik data on cosmic radiation has shown that both devices functioned normally. The dependence of the number of cosmic ray particles on the geomagnetic latitude was clearly discernable. Systematization of the great number of measurements of the energy spectrum of pure cosmic particles makes it possible to investigate the changes that come with time in this spectrum and compare them with the processes which took place at that time in the outer space around us.

Study of Biological Phenomena Under Conditions of Space Travel

To study several medical and biological questions, the sputnik had inside a special hermetically sealed chamber containing a test animal (a husky named Laika), measuring equipment to record the animal's physiological functions and also equipment for air regeneration, feeding, and removal of the animal's excretions. Equipment designs took into consideration the requirement for the most stringent economy in size and weight, coupled with a minimum consumption of electricity.

Functioning over a long period of time, the apparatus ensures by means of a radiotelemetering arrangement recording of the pulse beat and breathing of the animal, his arterial blood pressure and cardiac biopotentials, the temperature and air pressure inside the chamber, etc.

To regenerate the air inside the chamber and maintain the required gas composition, highly active chemical compounds were employed, giving off the necessary amount of oxygen for inhalation and absorbing carbon dioxide and the surplus of vapor. Their amount in chemical reactions was automatically controlled. Due to the absence of air convection under conditions of zero-g, an arrangement for automatic ventilation was installed in the animal's chamber. A set air temperature in the chamber was maintained by a heat-regulating arrangement. To give the animal food and water during flight, the container had special feeding devices.

The dog Laika, sent up aboard the sputnik, went through preliminary training. It was gradually accustomed to protracted periods of time in special clothing in small, hermetically sealed chambers, to the senders attached to its body in different places to record it physiological functions, and so on. The dog was also trained to withstand strain. The animal's resistance to the effects of vibration and some other factors was ascertained under laboratory conditions. After long training the animal was able to remain calm inside the hermetically sealed chamber over a period of several weeks, thus enabling the necessary scientific investigations to be carried out.

The study of biological phenomena involved in cosmic travel by living organisms was made possible through extensive advance experimentation with animals under conditions of short-term flights aboard rockets up to altitudes of between 100 and 200 kilometers (approximately 62 and 124 miles). These experiments were carried on in the USSR over a number of years.

At variance with earlier investigations, the animal's travel aboard the sputnik makes it possible to study protracted effects of zero-g. So far zero-g effects could be studied aboard aircraft, only within a a matter of a few seconds, or in vertical upward rocket flight, only in the matter of a few minutes. The sputnik makes it possible to study the condition of a living organism in a zero-g state lasting several days.

The experimental data derived through the fulfillment of the program of medical and biological research is currently being subjected to most detailed and thorough study. We can already say that the test animal stood up well to the protracted effects of acceleration when the sputnik was being aligned on orbit and to the subsequent state of zero-g which continued several days. The information received shows that the animal's condition was satisfactory throughout the entire experiment.

Without a doubt, these investigations will largely contribute to coming successful interplanetary voyages and will pave the way for evolving the means to guarantee safety to human beings in space travel.

◇ ◇ ◇

The launching in the Soviet Union of the first two sputniks makes an appreciable contribution to the study of the top layers of the atmosphere and extends the frontiers of human knowledge of the universe. At the same time this is evidence of our country's high scientific and technical standards, and allows us to foresee a day when all circumsolar space will be accessible to direct investigation by man.

Life In Sputnik

by P. ISAKOV, *M.Sc. (Biology)*

AN EXCEEDINGLY difficult, exceptionally important stage in mastering cosmic space has been attained. For the first time in the history of our planet man-made devices have penetrated interplanetary space and are revolving around the Earth. Following the first artificial satellite, which proved the possibility of systematic radio communication with cosmic space, another sputnik is circling the globe. It is much larger in size, equipped with instruments for research and even carrying a live passenger, the dog Laika. This was done in order to verify a number of theories concerning the action of many factors of cosmic space on living beings, to test the protective measures under the influence of factors which can prove harmful or fatal. The data received during this experiment are necessary for organizing future cosmic flights of people to the Moon, Mars and Venus.

Without Atmosphere

Flight of both animals and man in cosmic space is possible only in hermetically-sealed chambers in which the air, both in composition and pressure, is similar to that on Earth.

It is known that there is no air, no oxygen in interplanetary space.

To ensure the respiration of an organism in a cosmic apparatus a supply of oxygen is needed. For this purpose it is expedient to use liquid oxygen. During the evaporation of one liter of the liquid about 800 liters of gas are formed. But the use of liquid oxygen is practical for only relatively short flights. It would be difficult to take a huge supply of liquid oxygen for flights lasting several months or more. In such cases it would be more expedient to install hothouses with plants on the space ship as plants absorb carbon dioxide from the air and give off oxygen.

The absence of barometric pressure of the air is an important factor in interplanetary space as compared with the conditions of life on Earth. Let us recall that the necessary quantity of oxygen is soluble in blood only at definite barometric pressure. If the barometric pressure is insufficient, even the presence of pure oxygen cannot ensure the organism's being adequately supplied.

However, this does not exhaust the role of barometric pressure. All liquids in the organism, particularly blood, contain dissolved gases—oxygen, nitrogen, carbon dioxide, etc. If the barometric pressure of the air drops, the dissolved gases leave the blood. In case of a sharp drop in barometric pressure such a quantity of gases may escape at once from the liquids in the organism as will involve grave consequences and at times even full derangement of the organism's physiological functions.

The temperature at which liquids boil depends on the surrounding pressure. The lower the pressure, the lower the temperature at which boiling begins. At a pressure of the mercury column of 47 millimeters (corresponding to 19 kilometers above sea level), a liquid begins to boil at 37 degrees Centigrade—the temperature of human blood. The "boiling" of blood inevitably entails grave consequences.

When can such derangement occur? This can happen if the chamber is suddenly no longer airtight. Naturally, it is necessary to take precautionary measures to prevent such accidents.

How can this be done? Firstly, by creating the necessary barometric pressure in the chamber. But such a method is not very reliable, since any breach in the wall, anything making it less than airtight, would cause grave and often irreparable consequences. The second method is to use specially-designed clothing or space suits. The necessary barometric pressure is maintained in the space suit through the stretching of the suit fabric which grips the body tightly.

In practice the two methods are combined: barometric pressure is maintained in the chamber and special space clothing is used. The suit is brought into action when the chamber is no longer airtight.

The flight of animals in a sputnik will make it possible to ascertain how reliably airtight the chambers are, as well as the adequacy of the space suits, to work out the method of feeding and supplying water to organisms under these conditions and a number of other questions.

A suitable environmental temperature is an important condition for the normal existence of human beings. On Earth organisms are subjected to a fluctuation of the temperature of the environment in a relatively small range—approximately from $+70$ degrees to -70 degrees C. However, this does not mean that all animals are adapted to such fluctuations of temperature. Many can withstand only a considerably smaller temperature fluctuation. That is why thorough investigations have to be made of interplanetary space where temperature fluctuation can be immeasurably greater. The first experiment of launching a sputnik with a dog has already shown that Soviet scientists are correctly solving the problem of creating the necessary temperature regime within the sputnik. During the first hours of flight the dog behaved calmly and its general state was satisfactory.

The inhabited satellite is equipped with instruments for studying temperature and pressure.

Sun Means Not Only Life

It is exceedingly important to study with the help of sputniks the influence of various forms of solar and cosmic radiation on living organisms.

It is known that only a small part of the Sun's rays reach the surface of our planet. The rest is retained by the Earth's atmosphere. Thus, for example, the ultraviolet part of the solar spectrum is almost completely retained by the atmosphere. In the upper layers of the atmosphere, as beyond its bounds, the intensity of ultraviolet radiation is so considerable that it is fatal for living cells. But protection from the action of ultraviolet radiation from the Sun is not a difficult matter since most materials, including ordinary glass, keep out this part of the solar spectrum.

But solar radiation also has rays more unpleasant for living organisms. This is so-called X-ray radiation. The action of X-rays, at first entirely imperceptible to the organism, can lead to very undesirable consequences. That is why protection against them must be especially reliable.

There are, however, no grounds for fearing them excessively. On Earth, too, workers in some professions have to deal with such radiation. Protective measures have been worked out, reliably safeguarding people from them. There is no reason to assume that these problems will baffle designers of inhabitable space craft.

Cosmic rays, or cosmic particles, as they are more properly called, can be much more dangerous. They consist of nuclei of different chemical elements. Cosmic rays contain hydrogen nuclei to a great extent (about 80 per cent), with less of the heavier nuclei, nuclei of iron, for example.

The intensity of cosmic particles is not the same at different altitudes. Landing in the Earth's atmosphere, cosmic particles clash with air molecules, and their energy is lost in ionizing the molecules. Only an extremely insignificant number of cosmic rays penetrate the lower layers of the atmosphere. But at an altitude of about 200 kilometers their number increases 150-fold. The intensity of cosmic particles is still greater in interplanetary space. It is believed that beyond the bounds of the atmosphere the intensity of cosmic particles equals approximately 0.5 particle a second per one square centimeter.

The velocity of cosmic particles is exceedingly great, approaching the speed of light. Possessing tremendous kinetic energy, such particles clashing with molecules of other substances cause their dissociation and disintegration into ions. Ionization of molecules also takes place when cosmic particles penetrate the tissues of an organism, leading to the destruction of cells, to morbid phenomena similar to those caused by gamma-radiation arising from nuclear reaction.

The penetration of one particle into the tissue of an organism is still not dangerous. Even particles of such heavy elements as zinc and iron cause damage to only 15,000 cells approximately. Compared with the total number of cells in an organism, about 1,000,000 million, this of course is infinitesimally small. It is clear that the place where such a particle strikes is also of significance. It is one thing if a particle causes the destruction of fat cells in the subcutaneous layer and an entirely different matter in the case of the nerves of the cardiac muscle, or vitally important centers of the cerebrum.

The question, naturally, arises: How is the organism to be protected from cosmic particles? So far there are no fully completed projects on this score. The published data speak of great difficulty in protection. It is effected on the same principle as protection from nuclear radiation. The presence of nuclei of heavy elements in cosmic particles complicates this task. Even the most powerful artificial sources in laboratories have not yet created particles with energies which could be compared to the energy of heavy nuclei of cosmic radiation. The launching of a sputnik with animals will enable scientists to obtain highly important data on this question.

The task of science is not only to determine the amount of such penetration the tissues of an organism can withstand and to ascertain the intensity of cosmic particles, but also to establish the time in the course of which undesirable consequences may arise. It is known that the so-called latent period, i.e., the time from the moment cosmic particles enter the organism to the moment their action becomes evident, can last for weeks and even months. This necessitates prolonged observation of the organism after it has been subjected to the influence of the particles. For this purpose in a number of cases the animals, after staying in the sputnik, will have to be brought back to Earth for subsequent study.

The second Soviet artificial Earth satellite is equipped with measuring instruments for investigating all the above-mentioned "radiational" influences of cosmic space, short wave ultraviolet and X-ray radiation of the sun and cosmic rays.

It is necessary to say a few words about the so-called meteorite danger.

It is known that more than a ton of meteorites daily penetrates the atmosphere of our planet. Possessing huge velocities of 30-50 and more kilometers per second, they are heated by friction with the air and burn up in the upper layers of the atmosphere. As a rule, meteorites do not penetrate to altitudes lower than 70-100 kilometers. But the higher the altitude, the greater the possibility of a sputnik encountering meteorites. It is important to establish to what extent such a probability depends on the altitude of the orbit, the season of the year, etc.

The first weeks of flight by Sputnik I show it has not been subjected to the action of meteorites of destructive force. The meteorite particles which only make a hole in it are not a serious obstacle for ensuring the life of animals within the satellite. Protection against such particles, can be provided either by arranging a metal screen to absorb the energy of these particles at some distance from the shell of the sputnik, or by diverting these particles by electromagnetic radiation or by other means. It is not to be precluded that the probability of clashes with meteorite particles will prove to be no greater than that of motor traffic accidents.

How Much Does Laika Weigh in the Sputnik?

The living space in a sputnik is very rigidly limited, and for that reason the mobility of animals in a satellite has to be greatly restricted. Such restriction causes changes in the physiological functions of the organism. For example, we know that when a man is confined to bed for a long time it is often necessary to take measures to prevent hemostasis, bedsores, etc. Naturally, such measures have also been devised for animals within a sputnik.

A satellite has to be accelerated to about 8,000 meters per second. Since such speed has to be attained soon after launching, prolonged and considerable acceleration is necessary.

The effect of acceleration or overload, as it is called, on the organisms of animals and humans has been studied intensively in recent years. Large and prolonged acceleration occurs during flight on highspeed planes.

If acceleration acts vertically, from foot to head, then a redistribution of the entire mass of blood in the organism may result. There will be an overabundance in the lower part of the body and a shortage in the upper part. At the time of considerable intensity of acceleration the usual level of blood circulation in the cerebrum may be reduced, leading to a derangement of the functions of the central nervous system, even loss of consciousness.

If acceleration acts horizontally on man's body, it can be withstood much better. Special suits which tightly grip certain parts of the body and do not allow blood to accumulate in them help to fight the overload. These questions also are being studied in experiments with animals.

A few words about the speed of movement which an organism can stand. We know that uniform speed does not exert any effect on the organism. But the greater the speed of movement, the harder it becomes for man to orient himself in space. At a definite speed of movement there comes a time when man's sensory organs are unable to provide the brain with exhaustive information, in view of the rapid succession and incompleteness of the sensations. Therefore, the main control of cosmic flight will be from ground stations with the help of electronic computing machines in accordance with a program drawn up in advance. Thus, in cosmic flight astronauts will be free of directly controlling their flight, since this is literally beyond man's capability.

Let us proceed to the question of the absence of the force of gravity, or zero-g, which cosmic travelers will inevitably encounter. The action of zero-g has been studied both in the case of animals and man in recent years. Although these observations pertain to brief instances of zero-g, lasting no more than some seconds, there is no doubt that the influence of prolonged zero-g may prove to be entirely different. This is exactly what has to be established through the use of sputniks carrying animals.

Weight is a factor in all our movements. For example, in raising our arms we overcome the force of the Earth's gravity, while the same force facilitates movement when we lower arms. It may be assumed that if the force of the Earth's gravity ceases to exert its influence, the usual coordination of all motions will be upset. The muscular exertion of raising the arm, worked out in the course of the entire preceding life, may prove to be unnecessarily great. And conversely, the lowering of the arm will require unusually great exertion, since its weight which helps to lower the arm will be absent.

Such derangement in the coordination of motions has already been established in experiments with animals and man. At the same time these experiments have revealed the remarkable ability of the nervous system to adapt itself quite quickly to new conditions. After a certain number of experiments, coordination of motions under zero-g conditions begins to improve considerably.

Zero-g undoubtedly also influences such functions of the organism as respiration, blood circulation, body temperature, etc. We know that the weight of the blood is one of the factors in the regulation of blood circulation. The exclusion of this factor in zero-g conditions may affect the distribution of blood in separate parts of the organism. Observation shows that zero-g causes a certain lowering of blood pressure.

Under the effect of acceleration the gas exchange in the organism increases, the consumption of oxygen and the emission of carbon dioxide rise several times. During zero-g we may expect a lowering of the gas exchange, at least after the organism becomes adapted to the absence of gravity. These data are important for ensuring oxygen to animals and determining the capacities of air-conditioning devices. Naturally, experiments with animals within a sputnik will be quite fruitful in this respect as well.

There is one more important factor necessitating the launching of animal-carrying sputniks. We refer to the saving of a crew of future astronauts. We discussed earlier some protective measures under conditions of cosmic space when the chamber is no longer airtight. But unforeseen circumstances may compel the pilots to leave the space ships. It is necessary to foresee the possibility of saving the people in such cases as well. Naturally, it is expedient to conduct such experiment with animals at first. Moreover, for scientific purposes, also, it would be highly desirable to save the animals after a sputnik finishes traveling along its orbit. Specifically, this is necessary for studying the subsequent state of the animals.

Where Are the Biological Frontiers?

The distance of the orbit from the Earth is of significance for flights of sputniks without animals. The greater this distance, the longer the existence of the sputnik. What about animal-carrying sputniks? Is there any reason for seeking to reduce the distance of the orbit from the Earth? It turns out that at all altitudes where sputniks can be launched there is no difference as regards providing conditions for the life of animals. It makes no difference whether a sputnik travels within 300 or 1,000 kilometers from the earth. Equal conditions for the life of animals are necessary.

Beginning with very low altitudes, our atmosphere rapidly loses all the properties necessary for maintaining life. The ability of the atmosphere to maintain the normal gas exchange necessary for man extends only to an altitude of about five kilometers. Only a perfectly healthy man can go higher, and then only for a short time.

Insufficient barometric pressure of the atmosphere makes itself felt starting at an altitude of eight or nine kilometers. The appearance of so-called altitude pain, associated with the emission of nitrogen bubbles from the liquids of the organism, is possible at this altitude.

The so-called time reserve, that is, the period when man retains consciousness after the supply of oxygen is exhausted is practically the same, starting at an altitude of 19-20 kilometers. For example, if the supply of oxygen is cut at an altitude of 12 kilometers, a man can take the necessary measures without danger of losing consciousness for approximately 30 seconds. At 15 kilometers the time reserve is shortened to 15 seconds. At an altitude of 19 kilometers, as pointed out above, the time reserve is very small and hardly depends on the further increase of altitude.

At an altitude of 19,200 meters, where the general barometric pressure of the mercury column is 47 millimeters, it will be necessary to take measures for the protection of an organism from the boiling of liquids.

The borderline of absorption of heavy particles of cosmic radiation is at around 36-37 kilometers. Above that altitude serious protection from cosmic particles is necessary. The region where the ultraviolet part of the solar spectrum is equivalent to that in interplanetary space starts at 42-43 kilometers above sea level.

Meteorites usually burn up at an altitude of about 100 kilometers.

The propagation of sound becomes impossible at or above an altitude of 122 kilometers. At this altitude the distance between the air molecules equals approximately the length of sound waves man can hear. Above this border air molecules are even more scattered.

It is approximately at this altitude that the intensity of cosmic particles begins to rise sharply.

As for the region where sputniks can exist for a long time, it is much higher than all these boundaries.

Our times are times of rapid, ever-accelerating development of science and technology. The launching of sputniks, particularly of Sputnik II with an animal on it, signifies the advent of a new era in the history of science, in the history of man's mastery of cosmic space. At the same time it also signifies the beginning of a systematic advance into the universe. We believe that within the next five to ten years the flight of man into cosmic space, possibly with the landing on other planets of our solar system, will become a reality.

Komsomolskaya Pravda, November 5, 1957

Conducting Optical Observations

by Professor B. KUKARKIN, Ph.D.
Vice-President of the Astronomical Council
of the USSR Academy of Sciences

IN THE Soviet Union the organization of optical observations of the satellites' movements has been assigned to the Astronomical Council of the USSR Academy of Sciences, which has set up 66 special stations for this purpose in different parts of the country.

Each station has at its disposal up to 30 AT-I spyglasses, especially designed to observe the satellites. These instruments are small telescopes which have a magnification to the sixth power and have a large field of vision.

Shortly before the expected appearance of the sputnik the station sets up an "optical barrier"; the observers fix the spyglasses so that they cover the whole sector of the sky along a line perpendicular to the direction of the sputnik's movement. Often a second, supplementary optical barrier is set up, and the satellite will certainly be noticed as it crosses one of the barriers, and sometimes both.

What do the optical observations of the sputniks and carrier rocket tell us? We cannot, of course, arrive at final conclusions until after a thorough study of the many data obtained. But we can already sum up some preliminary findings.

An extensive program of scientific investigation is being carried out with the aid of the sputniks.

One of the objectives of observation with the aid of the sputniks is to study how intensively cosmic particles, including those from meteors, pass from outer space.

As the first sputnik moves along its path without any noticeable shifting, it may be concluded that it is not very much bombarded by meteoric particles.

After the launching of the second sputnik, which is much brighter, in addition to the special stations, aerological points of hydrometeorological service were enlisted to conduct visual observations. These have shown that the carrier rocket changes in brilliance, due to the change of its position in space. The shortest visually recorded period of change in brilliance is approximately 20 seconds.

Another thing which makes optical observation of the artificial satellites valuable is that they make it possible to determine the satellites' deviations from the theoretically traced path. The deviations can be caused either by considerable changes in the density of the upper atmosphere or by irregular distribution of masses in the interior of the Earth. The farther the path of the satellite is from the Earth's surface, the less it is influenced by the upper atmosphere. All the peculiarities of movement will in this case be caused chiefly by the irregular distribution of the masses in the interior of the Earth. Optical observations of the sputnik's movement will thus make it possible "to take a look" into the Earth's interior and study the distribution of the masses therein.

Along with visual observations, photographic observations of the sputniks and carrier rocket are being conducted. Photographs taken at the Pulkovo Observatory, the Observatory of the Astrophysical Institute of the Kazakh Academy of Sciences, the Observatory of Kharkov Univerity and other astronomical institutions in the Soviet Union, and by the Purple Mountain Observatory (People's Republic of China), the Edinburgh Observatory (Great Britain), the Potsdam Observatory (German Democratic Republic) and others have made it possible to establish more accurately the orbits of the sputniks and the carrier rocket.

In the Soviet Union and other countries optical observations of the sputniks are conducted by thousands of private citizens, in addition to the specialists. When conditions are favorable, the sputniks and the carrier rocket can be seen with the naked eye. Such conditions appear in the middle latitudes when the sputniks pass across the sky an hour to an hour and a half after sunset or at the same time before sunrise. Amateurs can also photograph them. All they need to do is to set the camera at Infinite, fix the camera so that the carrier rocket or sputnik will pass through the lens's field of vision. The instant the sputnik appears, the lens should be opened and remain open throughout the time the flight is being observed. The lens must not be diaphragmmed under any circumstance. While taking the photograph the time should be marked: After every five to ten seconds the lens should be covered with the palm of the hand for one second and the exact time recorded. The sputnik will leave a trace on the plate in the shape of a thin line with breaks corresponding to the intervals of the lens's obstruction. To make out the markings of time more easily the intervals between them should be made unequal.

It is best that two people do the photographing, one holding the watch and giving the signal, at which the other covers the lens. The most sensitive film should be used.

Radio observations and the recording of "bleeps" on magnetic tape also provide extensive material.

There can be no doubt that observations of the sputniks, classification and intensive study of the observations will help to pave the way for the era of the conquest of space, the era of interplanetary travel.

What Sputnik I Disclosed About Radiowave Propagation

Preliminary Report
by Professor A. KAZANTSEV
Doctor of Technical Sciences

THE LAUNCHING of the first Earth satellites has initiated a broad program of scientific exploration in the upper layers of the Earth's atmosphere and in the outer space surrounding our planet.

The reception of radio signals emitted by artificial satellites is one of the most important methods of this space exploration.

Radio-watch observations of Sputnik I were widely made at very many stations throughout the USSR and in other countries of the world.

The interpretation of the numerous reports received will, it stands to reason, take much time, and full results will be published later by appropriate organizations. It can be hoped that they will give important and interesting data on the structure of the ionosphere, its upper zones in particular, on absorption of radiowaves in the ionized layers and possible routes of the propagation of short radio waves.

Preliminary results of reception of Sputnik I radio signals show that on the 15-meter wave these signals were received at very great distances, far surpassing the distance of direct visibility and in a number of cases reaching 10,000 kilometers. Study of the data on long-distance reception of these signals will undoubtedly yield very valuable material on possible routes of shortwave propagation.

But what are the conditions of propagation of short waves emitted by Sputnik I?

As *Pravda* reported in the lead article "Soviet Artificial Earth Satellite," the perigee of the satellite orbit (its lowest point) is in the northern hemisphere, and the apogee (orbit's highest point) is in the southern hemisphere. The altitude of the apogee above the Earth's

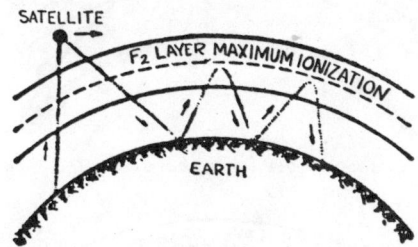

Fig. 31 Satellite above ionosphere (direct, reflected reception)

surface reaches approximately 1000 kilometers. In the southern hemisphere, therefore, the satellite travels above the principal layer of the ionosphere which conditions the reflection of short radiowaves, layer F_2.

As far as the northern hemisphere is concerned, especially interesting conditions of shortwave propagation are created in it. At certain intervals of time Sputnik I was above the maximum of ionization of layer F_2, at some intervals below it and at certain times close to this maximum.

When Sputnik I was above layer F_2, then the radiowaves passing from above through the whole thickness of the ionosphere, were reflected from the Earth's surface and propagated further through single or multiple reflection from layer F_2 in those areas where its critical frequency has sufficiently high values (see Fig. 31).

It is also possible that radio waves falling on the ionosphere from above at a sloping angle undergo in it considerable refraction, and owing to this penetrate into a zone lying beyond the limits of direct geometric visibility (Fig. 32).

When Sputnik I was below the maximum of ionization of layer F_2 (Fig. 33) and approached a point of observation from an area of the globe illuminated by the Sun, the radio signals on the 15-meter wave could come from it to a point of reception, at first after experiencing consecutive reflections from layer F_2 and the Earth's surface, and then already by virtue of direct visibility.

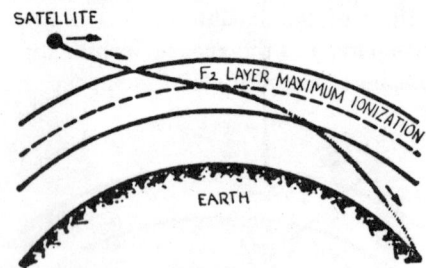

Fig. 32 Satellite above ionosphere (slanted reception)

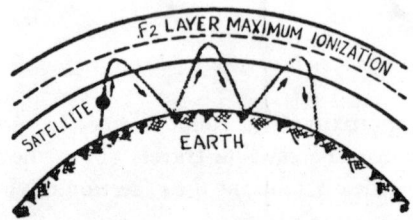

Fig. 33 Satellite below ionosphere (reflected reception)

If passing over the district of observation, the satellite then moved away into an area of the globe not lighted by sunshine, then reception of signals ceased in a comparatively short distance, determined by the limits of direct visibility.

Non-symmetrical conditions of reception were also observed in a number of cases.

When the satellite was close to the maximum of ionization of layer F_2, then especially favorable conditions might be created for the formation of radiowave conducting channels capable of propagating radiowaves for very long distances (Fig. 34).

In fact, there are indications that along with satellite signals which had arrived at the observation point by the shortest distance, signals were sometimes received which had travelled around the globe (round-the-world radio echo).

One of the most experienced shortwave radio amateurs in the USSR, Yu. N. Prozorovsky (Moscow) on October 8 at 0007-0008 hours recorded the reception of such a round-the-world radio echo on the 15-meter wave length.

Concerning the signals on the 7.5 meter wave length, as far as can be judged from available data, they were as a rule received in the limits of direct visibility, although in separate cases owing to

Fig. 34 Satellite near F_2 Layer Maximum Ionization (round-the-world echo)

high values of critical frequencies of F_2 layer in the daytime, this wave could be propagated also beyond the limits of direct visibility, according to the law of shortwave propagation.

After having established the correlation between the alititudes at which Sputnik I was and the real altitudes of the maximum of ionization of layer F_2 at one and the same moment of time, and then analyzed the conditions of radiowave propagation at that time, a conclusion can be drawn as to precisely what route radiowave propagation followed.

Radio, No. 12, 1957
Moscow

Russian Air Force Views On Sputniks

THE SOVIET air force played the leading role in launching Sputniks I and II. What aeronautical experts in the USSR think about Earth satellites was the theme of a special issue of *Sovietskaya Aviatsiya* (Soviet Aviation), the newspaper of the nation's military air forces, on December 1, 1957. Selected from this issue are the four short articles appearing below.

1. Rocket-plane, the Aircraft of the Future

by V. ALEKSANDROV

Candidate of Technical Sciences

The swift advance of rocket technology has now opened up the possibility of creating an aircraft that can fly in the upper reaches of the atmosphere. Such supersonic planes of the outer atmosphere will be called rocket-planes.

A rocket-plane, or rocket-powered glider, is a flying machine that can bolt like a rocket to outer space and on return to Earth land like an ordinary airplane. After taking off from Earth, the plane skyrockets up to get as swiftly as possible into rarefied layers of the atmosphere. When flight speed reaches 12,000 to 15,000 kilometers an hour, the pilot switches off the rocket engines, and the plane continues in coasting flight on its own momentum like an artillery shell. It describes an elliptical curve (see Fig. 35) and descends to dense layers of the atmosphere. Here its wings begin to create the necessary lift, and it shifts over to gliding flight. Bearing down on the air and utilizing its vast speed resources, the flying vehicle skips up again into the atmosphere's rarefied layers and describes a second, and additional ballistic curves until its entire reserve of kinetic energy

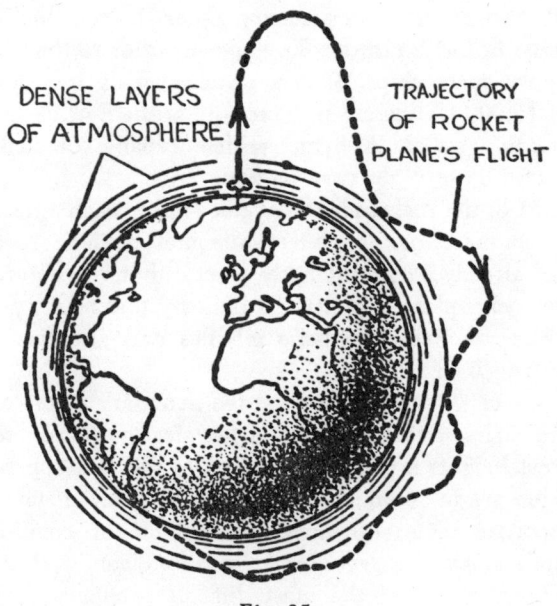

Fig. 35

has been exhausted. A rocket-plane can thus fly for many thousands of kilometers without consuming any fuel. Its flight trajectory will resemble the skipping in water of a flat stone skilfully thrown at an angle onto a still water surface.

But how will the rocket-plane land on Earth?

The rocket-plane's speed must obviously be slowed down, to prevent its being burned up on reaching the dense layers of air. The best braking method is rapid increase of the air friction area. We call to mind, for example, how the eagle behaves when approaching the

Earth. With folded wings at first, the bird falls earthward like a stone. But when near enough to the ground, the eagle opens its wings full sweep, and parachuting lands on Earth without impact. The rocket-plane has likewise to "open its wings," i.e. abruptly to enlarge its resistance area, when it is still high above the earth at an altitude of 80 to 100 kilometers. Only in this case can the plane rapidly slow down from its supersonic speed and make a satisfactory landing.

It is clearly a complex technical problem to create a flying machine which on take-off will have the properties of a rocket-ship and on landing the properties of an ordinary glider. Indeed this is virtually the last stage in Earth aircraft development prior to the transition to interplanetary space ships. Such a plane must fly in outer space at a speed of 15,000 kilometers an hour, and at the Earth's surface have normal landing speed. A flying vehicle capable of such variable speed must while in flight vary its wing area.

The thrust of the rocket-plane's engines must reach at least 50 tons. The total engine thrust of intercontinental ballistic rockets, it is known, can already be considerably greater than this figure. For this reason the rocket-plane will first be built by that country which has the best intercontinental ballistic missiles and also has new high-output heating fuels.

Rocket-planes will fly not only on the boundary of the atmosphere but also in airless outer space. It is therefore necessary to consider that at great heights a considerable part of the atmosphere's oxygen and nitrogen are in atomic form, and also in ionized state, and have other properties. This will have an effect on the condition of the atmosphere's boundary layer and its aerodynamic characteristics. Involved here is one of the most difficult problems in creating a rocket-plane, that of methods for piloting it. To change a rocket-plane's flight direction is incomparably more difficult than it is to guide an ordinary airplane flying near the ground. In changing its flight direction, the airplane can have bearing on the air. At altitudes above 200 kilometers such bearing is practically non-existent. One means of changing a rocket-plane's flight route can be the use of auxiliary rocket engines, set at an angle to the flight direction. The thrust of such engines will change the travel trajectory in the desired direction.

Another problem not yet completely solved is the creation of necessary living conditions for the crew and passengers under the impact of immense accelerations. At take-off, rocket-plane passengers must not experience gravity overload strains of more than five G.

When travelling on the descent branch of the rocket-plane's ballistic curve, its passengers will be in conditions of zero gravity. How prolonged weightlessness will affect the human organism and the rate at which the pilot recovers orientation, we are still unable to say exactly. That is why scientists and designers still have much research work to do in designing rocket-planes capable of flying with a speed of 10,000 to 15,000 kilometers an hour at altitudes reaching 1000 kilometers.

The epoch of rocket-planes is brought considerably nearer by the creation in our country of powerful intercontinental rockets and the successful launching of two artificial earth satellites.

Space ships, like the rocket-plane, will become mighty means for developing the economy and culture of mankind.

2. Intercontinental Ballistic Missiles and Aviation

Professor G. I. POKROVSKY, *Doctor of Technical Sciences*
General-Major of Engineering-technical Service

The creation of an intercontinental rocket and its use for launching two artificial earth satellites is a very great victory for the science and technology of our Socialist fatherland. Representatives of Soviet aviation are especially glad to acknowledge this. Indeed the mighty new rocket technology grew up predominantly on the basis of Soviet aviation culture.

The combat power of intercontinental missiles makes it possible, in case of necessity, to strike swiftly and accurately with hydrogen bombs any military targets in any part of the globe. This substantially enlarges the possibilities of modern military technology, singularly strengthens our armed forces.

It can, at the same time, be said with confidence that further development of space flights will bring aviation and rocket technology into an original closer association. The problem of creating a space ship that will return to Earth can be solved, for example, only by giving this ship wings and controls peculiar to modern airplanes. The rocket ship will need wings for effective braking of its flight speed in the atmosphere and making a landing at a given airdrome.

The airplane form in the design of rocket-carriers for a space ship is also expedient, as it will allow for their return to an airdrome for repeated use. The more cosmic voyages there are, the more important multiple use of rocket-carriers becomes.

Aviation and rocket technology are thus close and related fields of advanced Soviet science, and their development is closely interrelated today already.

3. Heart of the Rocket

by V. GRENIN, *Engineer-major*

Who among us has not admired the swift flight of jet planes, passing like lightning over the horizon or tracing white vapor trails across the blue sky! The supersonic speeds of modern aircraft became possible with development of their powerful air-breathing jet engines.

But air-breathing jet engines are useless for flights in outer space. The air they require is absent outside the earth's atmosphere.

Only rockets carrying a supply of oxygen on board can make flights beyond atmospheric limits. Their engines operate without air. A rocket's flight distance and speed is determined first of all by the power and faultless performance of its engines. The engine is therefore called the rocket's heart.

But how is this "heart" organized?

In modern rockets two kinds of engine can be used—the liquid-jet fuel and the solid fuel (gunpowder) engines. The liquid-fuel engine is widely used.

The design of a liquid-jet engine is rather simple (see Fig. 36). It has three main parts: the head, combustion chamber and nozzle. Placed in the head are the spray burners which atomize the oxidizer and fuel supplied from tanks by pipeline.

The mixing of fuel components, their ignition and combustion takes place in the combustion chamber which has the form of a cylinder or sphere.

The combustion process in the liquid-fuel rocket is of high intensity. Several score atmospheres of pressure develop inside the chamber. Heat and gaseous combustion products are formed each second in immense quantity. Gas temperature in the jet exceeds 3,000° C.

At such temperature many metals melt. How prevent burning-up of the engine? The walls of the combustion chamber and jet nozzle are liquid-cooled on the outside by the fuel which is circulated in the space between the engine's inside walls and outer jacket. Circulating at a definite flow rate, the fuel cools chamber and nozzle walls, and then through the spray burners enters the engine already preheated.

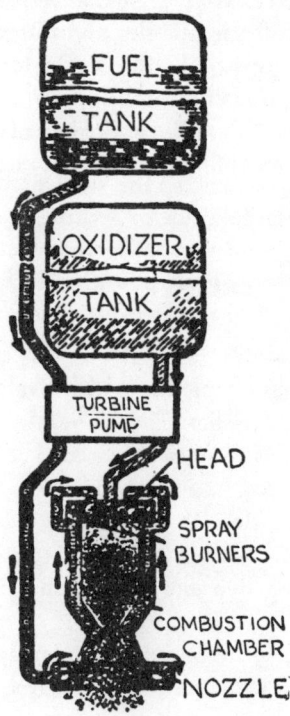

Fig. 36

Preheating of the fuel has a favorable effect on the fuel burning process.

Many tons of fuel are burned each minute in the engine. For fuel delivery to combustion chambers, a highly efficient reliable feeding mechanism is required. In large rockets fuel components are fed from tanks to the engine by powerful centrifugal pumps. They are actuated by a special turbine and are built as one unit with it, called the turbine pump.

The oxidizer and fuel mixed together in the combustion chamber form powerful explosive mixtures that are extremely sensitive to various outside actions (impacts, sparks, detonations). Therefore, the least disturbance in the feeding of fuel components or variation of their ratios or delay in ignition can cause an explosion and wreck the whole rocket.

The liquid-fuel rockets are the most effective kind for use in modern intercontinental ballistic missiles and future space ships.

Soviet designers were first in the world to develop powerful and reliable engines for intercontinental ballistic missiles and for rocket-carriers of artificial earth satellites. Soviet success in rocket-building can to a considerable extent be attributed to achievements in developing first-class engines.

4. Ships of Interplanetary Space

by V. KAZNEVSKY, *Engineer-designer*

From time immemorial people have with lively interest observed the movements of heavenly bodies. The Moon, the stars, the Milky Way's silvery strip and "shooting stars" have always attracted the attention of mankind and inspired human thought about the structure of the universe. From generation to generation man has passed on the aspiration to know the secrets of the mysterious stellar spaces, the natural laws governing live and inert matter in outer space beyond the planet Earth.

If these secrets are to be bared and interplanetary space explored in detail, man armed with the latest research tools will necessarily have to penetrate into space himself.

The space rocket is at present generally recognized as the sole and irreplaceable means of flight in outer space. The rocket has an inestimable quality; for travel it does not require outside bearing or carrier and is able to fly by the backward thrust of gases ejected from its jet engine's combustion chamber.

Ordinary air flight on Earth will be superseded by specific interplanetary travel, with its astonishing swiftness, smoothness, noiselessness and extremely diminished weight, bordering on complete loss of gravity. Millions of kilometers in distance will be covered at speeds reaching 70,000 to 100,000 kilometers an hour.

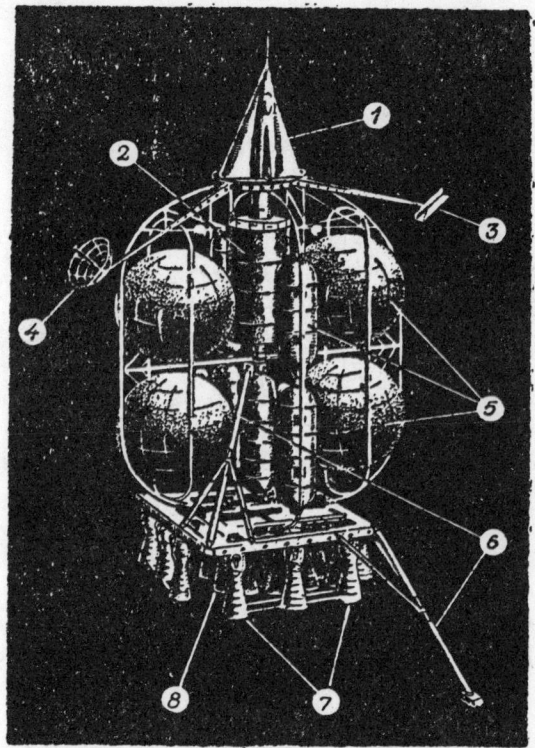

Fig. 37 DIAGRAM OF A SPACE SHIP DESIGN
1—"*retro-rocket*"
2—*passenger cabins*
3—*solar mirror*
4—*directional antenna*
5—*fuel tanks*
6—*landing gear*
7—*mobile engines*
8—*stationary engines*

But what will the interplanetary space ship look like?

The space ship will have an unusual form, unlike the streamlined aircraft we are now accustomed to. It will be a series of spherical and cylindrical hulls, with large, convenient and spacious rooms for astronauts (see Fig. 37). Space ship cabins will be more roomy than airplane cabins. Trips will take longer. On a flight to Mars, for instance, the astronaut will have to spend about one year in the ship.

The interplanetary ship will have a nuclear-powered jet engine.

Light, strong alloys, highly resistant to extremes of heat and cold, must serve as building material for the space ship.

The space ship must have a solar generator of electric power, reliable radio equipment, modern radar units and automatic controls to keep the vehicle on a pre-set trajectory and at the desired flight speed.

Artificial satellites are to be used for launching space ships on flights to other planets. Such a space flight starting point has advantages. First, the Earth's gravitational pull has been partially overcome on an artificial satellite and to escape from it entirely the space ship's speed has only to be raised about 3 kilometers per second. Secondly, future space ships will evidently be powered by a nuclear or photonic (light energy) engines. The launching of such rocket ships from the Earth's surface can have undesirable consequences in view of their dangerous radioactive exhaust. But a take-off from an artificial Earth satellite would avoid the hazardous radioactive fallout effects on the Earth's surface and atmosphere.

An interplanetary ship can start from a satellite with quite small accelerations and gradually pick up speed, which will make the work of astronauts easier.

After an artificial satellite has been orbited near the Earth to form a transfer station and necessary parts have been delivered there, the space voyagers can assemble their interplanetary ship and set forth on a trip to other planets. When approaching a planet, the ship will brake its flight speed sufficiently to become itself an artificial satellite of the planet.

A space landing craft in the form of a small "rocket-boat" or "retro-rocket" will then detach itself from the ship, taking astronauts down for a "soft" landing on the planet.

For the return trip to Earth, a rocket speed sufficient for the take-off from the planet will be necessary. The flight speed of 8 kilometers per second already reached by present Earth satellites is adequate for taking-off from the Moon or Mars on a return to Earth.

The rapid modern development of jet propulsion inspires us with the hope that space ship flights to the planets nearest the Earth will become possible in a few years' time.

Man-made Sun

Space flight promoters in the USSR are obviously planning an unmanned flight to the Moon as the next step in their astronautical program. Meanwhile, there is talk about creating a man-made thermonuclear Sun for the Earth.

"Soviet scientists are at present successfully solving the problem of the possibility of realizing controlled thermonuclear reactions," writes Academician V. S. Kulebakin in this same issue of the air force newspaper *Sovietskaya Aviatsiya*. "A series of interesting theoretical and experimental results have already been achieved in this direction. The realization of controlled thermonuclear reaction will permit us to light our own artificial Sun on Earth."

Scientist Kulebakin did not elaborate on this intriguing idea suggested in a review of space age prospects confronting mankind.

Soviet Sputniks And Radio Electronics

by Academician A. I. BERG

SPECIALISTS in jet propulsion and rocket engines played the paramount role in launching the Earth's first artificial satellites. But of no less importance was the successful work of radio electronics experts who created the instruments for guiding the rockets, putting the sputniks in planned orbits and providing for transmission of radio signals to Earth.

After satellites have been orbited, the space exploration program is in the hands of radio specialists, astronomers and mathematicians. Specially significant becomes the work of calculators and theoreticians armed with electronic computers. They process the results of observations made with optical and radiotechnical instruments.

Radio observations of earth satellites are extremely important and very extensive. How extensive can be judged from the fact that shortwave and ultra-shortwave signals from the satellites were received by more than 300 radio watch stations in the most varied districts of the Soviet Union. To the address "Moskva-Sputnik" came thousands of reports from Khabarovsk and Leningrad, Magadan and Kaliningrad, Archangelsk and Tashkent, Viln'ius and Tbilisi, L'vov and Alma-Ata and other cities of the country.

Satellite radio signals were at many places recorded on magnetic tape; field intensity was measured; the period of audibility and the character of the radio signals were systematically registered. Radio amateurs played a large role, doing a job that could not have been done otherwise except by a virtual army of specially-delegated scientists. Among radio amateurs who tracked the satellite were many from foreign countries, including Poland, Czechoslovakia, East Germany, Italy, the United States, the Union of South Africa, the Azores and Israel.

The data collected in the satellite watch observations will serve to help organize coming space flights to the Moon.

The data pickup units and special telemetering instruments installed in the second satellite, a kind of space ship observatory, transmitted to Earth for seven days data on temperature and pressure in upper atmospheric layers, information on cosmic radiation intensity, on ultraviolet and X-ray radiation of the Sun, on the pulse, respiration, heart activity and state of health of the test dog Laika.

Of special interest for radio specialists was the data picked up on solar radiation in the shortwave band which has a direct effect on conditions in the upper layers of the atmosphere.

For more than a hundred years scientists have been exploring the intensity and spectral composition of the radiant energy which falls on the Earth from the Sun, and have on this basis indirectly been attempting to determine what these magnitudes are for conditions outside the Earth's atmosphere.

The most reliable data at present permit assuming that the density of the stream of the Sun's radiant energy, beyond the limits of the atmosphere, is equal to 1.4 kilowatt per square meter. In actinometry and meteorology this magnitude is called the "solar constant." About 9 per cent of this stream falls on the ultraviolet part of the solar spectrum, about 40 per cent on the visible part and 51 per cent on the far red and infrared of the Sun's spectrum.

At the Earth's surface, with the Sun standing at an altitude of 30° above the horizon, the density of the stream of solar energy is considerably less owing to the dispersion and absorption of solar energy by the atmosphere; it amounts to not more than 30 to 35 per cent of the stream density beyond atmospheric limits and is differently distributed. Only 2 to 3 per cent of it falls in the spectrum's ultraviolet part, 44 per cent in visible spectrum and 54 per cent in spectral heat rays.

The making of these data more precise, particularly the direct measurement of the stream density of the Sun's radiant energy, i.e. the solar constant beyond the limits of the atmosphere, will make it possible to determine accurately the Sun's effective temperature and the density of the radiant energy stream emitted by a unit of solar surface. Precise measurement here is of interest to astrophysics first of all, but it has more than cognitive importance.

If a transistor solar battery of 1 m^2 in area be constructed and faced toward the Sun even with the accuracy of a 30 degree angle, then as might be expected this surface will be exposed to solar power of about 1 kilowatt. With 10 per cent battery efficiency in conversion of solar energy to electricity, the output from such a solar battery surface might be expected to reach 100 watts of electric power.

But if it be assumed that a satellite flying at a great height is exposed to the rays of the Sun approximately two-thirds of its orbit circuit time around the Earth, then the solar battery can be expected to produce 100 watt-hours of energy. However, to secure such conditions, the spectral characteristic of the transistor battery must be close to the above-indicated frequency distribution of solar energy, especially in the visible and infrared parts of the spectrum, and moreover, such a battery must operate on an optimal load.

Unfortunately, the materials presently known which will permit creating batteries with high internal resistance are complex and cumbersome. A much lower magnitude of electric energy should therefore be expected. But even this would nevertheless have great importance as a possible alternate source of power for satellite measuring instruments; for example, a solar battery used in combination with an ordinary or a storage battery.

Of great importance are observations made on characteristics of the propagation of radio waves emitted by the satellites. The main source of information about the ionosphere has heretofore been the study of radio waves sent from the Earth and reflected back from those areas of the ionosphere which lie below the zone of maximal ionization. Therefore, the precise altitude of the ionosphere's upper limit was unknown. To establish this limit, it was necessary to get signals that had entered the ionosphere from above, from outer space. Sputnik I became just such an out-riding correspondent. Its signals came from those areas of the ionosphere that had previously been inaccessible for prolonged observations.

Measurement of received signal levels, and the refraction angles of radio waves sent at differing frequencies, makes it possible to get data on radio wave attenuation in those ionospheric areas which had not been investigated before, and in addition to collect certain data on the structure of these areas.

Not less importance is attached to direct measurement of the upper atmosphere's air composition, its pressure and ionic composition, also to measurement of the Earth's magnetic and electrostatic field, data on cosmic rays, meteoric dust and other phenomena. While of great practical value, the usual measurements made by numerous ionospheric stations leave many important problems unsolved.

It is obvious that without radio signals coming from the satellites, their flights would be of substantially less value, since with their comparatively small sizes, their vast orbital distances and altitudes, only episodic optical observations of them would be possible when the satellite passes relatively near the observing post and the weather is good. Even use of such a modern means as radar observation can only slightly improve the situation in the absence of satellite radio signals.

Radio physics and radio technology no less than jet propulsion must develop to a high level for solution of all the above-listed problems of science. Only accurate and reliable instruments can carry out the numerous measurements over a prolonged period. But because of the small space and dimensions of satellites, it is extremely difficult to install essential satellite instruments and power supply.

Such intricate and "intelligent" instruments, it goes without saying, cannot be one of a kind. The choice of measurement methods, instrument designs, their manufacture and reliable built-in accuracy in coordinated operation require high level technical knowledge and the know-how of making very complex instruments on a large scale.

Earth satellites and their automatic instruments characterize the level of a country's science and technology, and also, what is especially important, the nation's capacity to mobilize groups of specialists in highly diversified fields for combined action to solve a single common task.

The palm of priority in launching the earth's first artificial satellites will always belong to the USSR. Hardly anyone in the USSR, however, thinks that American scientists and engineers are, therefore, less talented than Soviet space flight experts are.

Some people in America apply old yardsticks to measure the potentialities of Soviet science and spread fairy tales about Russia. It is precisely for this reason that the Russian sputniks struck such a blow to American psychology and "shocked" public opinion so much in America.

Soviet radio physicists, radio engineers, designers and radio amateurs know perfectly well that, while they have done much, a great deal more remains to be accomplished. The results achieved, however, give solid grounds to hope that the more difficult problems we now confront will also be solved with success.

Radio, No. 1, 1958
Moscow

What Sputniks I and II Disclosed About Outer Space

IN CONFORMITY with the International Geophysical Year program, a third artificial Earth satellite was launched in the Soviet Union on May 15, 1958. Designed for the continued study of the upper layers of the atmosphere and cosmic space, it is equipped as a scientific laboratory with a much greater variety of instruments, making it possible to obtain new, valuable information. These instruments occupy approximately three cubic meters and have a combined weight of 968 kilograms. Carrying out the entire research program through the launching of sputniks will enable scientists to determine the structure of the upper layers of the atmosphere, the important processes going on in them and to proceed in elaborating a general theory of basic atmospheric phenomena. Below we publish some results of scientific investigations conducted with the help of the first two sputniks, launched in October and November of 1957, as recently published in the Soviet press:

The successful launching of the first artificial Earth satellites signifies the beginning of man's penetration into outer space. The sputniks open up the broadest prospects for many highly important scientific investigations. Of practical interest are the study of the ionosphere and the mechanism of its formation; the action of the Sun's radiation and cosmic rays on the Earth's atmosphere; the study of density, temperature, magnetic and electrostatic fields at high altitudes, etc. The solution of these problems requires direct experiments at altitudes hundreds of thousands of kilometers above the surface of our planet. The possibility of conducting such experiments has become a reality with the creation of the sputniks, enabling scientists to make the necessary measurements at high altitudes over various areas of the globe and for a long time.

Although the significance of artificial Earth satellites for research was known long ago, their launching was an insoluble problem until recently. The main difficulty was to develop a rocket capable of imparting to a sputnik a speed of approximately 8,800 meters per second. Only after an intercontinental ballistic rocket had been produced in the Soviet Union was an artificial earth satellite launched successfully for the first time. The superior design of this rocket made it possible to put into orbit sputniks carrying scientific instruments of considerable weight. Sputnik I weighed 83.6 kilograms, while the scientific and measuring instruments and sources of power on Sputnik II weighed 508.3 kilograms.

The launching of sputniks with instruments of such weight made it possible to carry out a whole range of scientific investigations: the fact that these studies could be conducted simultaneously greatly enhances their scientific value. Only through the development of large sputniks can the problem of creating regularly operating cosmic laboratories and making interplanetary flights be solved.

The scientific tasks set in launching the first sputniks determined the parameters of their orbits. Sputnik I was put into orbit with a perigee of 228 kilometers and an apogee of 947 kilometers. For Sputnik II the figures were 225 kilometers and 1,671 kilometers respectively. The period of revolution around the Earth at the start was 96.17 minutes for sputnik I and 103.75 minutes for Sputnik II. As the sputniks moved along their orbits, it was possible to make a number of experiments in studying the upper atmosphere (determination of the density of the atmosphere, study of the propagation of radiowaves, etc.). On the other hand, at these altitudes the density of the atmosphere is sufficiently low and therefore does not distort the measurements of the primary component of cosmic radiation, the spectrum of shortwave radiation of the Sun, etc.

Scientific tasks also determined the angle of incline of the orbit to the plane of the Earth's equator, equaling approximately 65 degrees. The advantage of such an orbit lies in the fact that in the course of the sputnik's flight the scientific instruments it carries can take measurements over different latitudes. It should be noted that to place a satellite in orbit with a greater angle of incline to the plane of the equator is a much more difficult task than to place it in orbit close to the equator.

Sputnik I made some 1,400 revolutions around the Earth during its existence from October 4, 1957, to January 4, 1958. Sputnik II made some 2,370 revolutions from November 3, 1957, to April 14, 1958.

The outlined program of research was carried out successfully with the help of the first two Soviet sputniks. Some preliminary results of these studies are presented below. On the whole the accumulated data are quite extensive and their analysis is continuing.

Radiotechnical and Optical Observations of the Sputniks

Since an analysis of the changes in a sputnik's orbit as regards time makes it possible to estimate the density of the upper layers of the atmosphere, studies of the movements of sputniks are of great significance. The elements of a sputnik's orbit can be determined by tracking it by radiotechnical and optical methods.

The radiotechnical methods included radio direction-finding and observations of the Doppler effect during the reception of radio signals from the sputniks. The Doppler effect is a result of the fact that the frequency of signals received increases as the object on which the radio transmitter is installed draws nearer to the receiving point. The changes in the frequency depend on the speed at which the object draws nearer or moves away. In the case of a sputnik the speed at which it draws nearer to, or moves away from the receiving station on the ground is so great that the Doppler effect can not only be observed on an ordinary radio set, but can also be used for registering the moment the sputnik passes at the distance closest from the point of observation and also for measuring the distance to the sputnik and its velocity.

During radio observations of the signals of Sputniks I and II, the frequencies of the signals received were measured by special radio equipment, including a recording chronograph.

To obtain greater accuracy of measurement observations were

conducted of signals at frequencies of 40 megacycles per second, which are less subject to the influence of the ionosphere. The power of the transmitters ensured the definite reception of the signals within the entire zone of direct visibility. Six or seven passages of the sputnik over the ground stations could be observed consecutively during the course of 24 hours.

To analyze the radio signals received a method was worked out making it possible to determine with a precision of 0.1-0.2 seconds the moment at which the sputnik passes at the shortest distance from the observation point.

The observations have confirmed that the Doppler effect can successfully be employed for determining the parameters of the sputnik's orbit. Simplicity and dependability of the equipment are distinctive features of this method. By raising the frequency of the transmitter installed on a sputnik and by automatic registration of the frequencies the errors of this method can be substantially reduced.

The most simple method of optically tracking the sputniks was by registering the moment of their passage over the observation point.

For a more precise determination of the bearings special methods were employed; modernized aerial cameras were used for obtaining photographs of the track of the sputnik. The time during the filming was marked by several consecutive openings and closings of the shutter, and the timing registered by a photoelectric method. In this way a distinct track of the sputnik was visible on the photograph. A high degree of precision was obtained when using such cameras.

A method of photography with highly-sensitive equipment has been worked out in tracking the artificial satellites. Very promising among these are electronic-optical transformers. The new method makes it possible to track the sputniks without the use of large optical systems, greatly simplifying the equipment necessary for observation.

Determination of the Density of the Atmosphere

The density and temperature of the air are the most important characteristics of the atmosphere. Their determination at high altitudes up to the limit of the atmosphere is essential for understanding a number of geophysical phenomena. For example, temperature influences the degree of ionization of the atmosphere, which in turn affects the propagation of radiowaves. The movement of meteorites and the stream of minute particles in the atmosphere depends on its density. The swiftest atoms and molecules on the edge of the atmosphere escape beyond its bounds and fly into interplanetary space. The speed of this process depends on the temperature at high altitudes. Lastly, in launching artificial satellites, it is necessary to know how long they will exist, for which purpose data on the density of the atmosphere are needed.

Conceptions about the upper atmosphere have originated on the basis of indirect data (observations of the aurora borealis, meteorites, etc.). These observations led to conclusions about relatively high values of density and temperatures. Later on, as a result of an analysis of investigations conducted with the help of rockets in recent years and a number of theoretical considerations, another viewpoint has gained general acceptance that the upper atmosphere is colder and less dense than previously supposed.

Even prior to launching the first artificial satellites scientists pointed out that observations of their movements offered the possibility of determining the density and temperature of the atmosphere. As they travel in the atmosphere, sputniks encounter resistance. This force of resistance is proportional to the density of the atmosphere. As a result of the decelerating influence of the atmosphere, the height of a sputnik's orbit is gradually lowered. This process continues until the sputnik, landing in the denser layers of the atmosphere, ceases to exist.

The density of the atmosphere declines sharply as the distance from the earth's surface increases. That is why the force of resistance is unequal at different sections of an elliptical orbit. With a sufficiently elongated orbit the force of resistance in the perigee is much greater than in the apogee. Hence, the main deceleration takes place in the area of the perigee. Such a nature of varying deceleration results in the height of the apogee declining much faster than that of the perigee. A sputnik's elongated orbit changes so that its shape gradually becomes more of a circle.

After the launching of the first sputniks optical observations and radio tracking made it possible to trace the evolution of their orbits. Since the action of the atmosphere on the sputnik on separate sections of its orbit is very small, scientists so far have not succeeded in measuring local deceleration. All data of the orbits immediately after the launching of the sputniks and also the changes in the periods from revolution to revolution throughout their life were measured on the basis of the observations of the first Soviet Earth satellites with a precision sufficient for definitely determining the density of the atmosphere.

The speed of change in the period of revolution greatly depends both on the density of the atmosphere in the perigee area and also on the speed with which the density declines as the altitude increases. The speed in the drop of the density is characterized by a parameter called the "height of uniform atmosphere," which is directly proportional to the temperature of the atmosphere and inversely proportional to its molecular weight.

On the basis of a theoretical analysis of the results of the observations scientists succeeded in definitely determining the value of the product of the density of the atmosphere and the square root of the "height of the uniform atmosphere" at altitudes of the perigees of the first sputniks (225-228 kilometers). The values of the density were calculated for definite theories regarding the value of the "height of a uniform atmosphere." The obtained value of density proved to be five to ten times greater than the values of density at these altitudes indicated in a number of models of the atmosphere built on the basis of rocket measurements prior to the launching of the sputniks. It should be noted that determination of the density by a study of the purely mechanical action of the atmosphere on the sputnik is quite exact.

The atmosphere is not the same over various areas of the Earth's surface. At the same altitudes the density and temperature change, depending on the latitude and time of day, which in turn is related to the unequal heating of the upper atmosphere by ultraviolet, X-ray and minute-particle radiations of the Sun.

As a result of the fact that the gravitational field of the Earth differs from the central one, the orbits of the sputniks changed their position in space. Thus, for the first Soviet sputniks the angular distance of the perigee from the midday meridian changed approximately by 4 degrees and the latitude of the perigee changed by 0.35 degrees in 24 hours.

Inasmuch as the main action of the atmosphere occurs in the

perigee area of the orbit, the change of its position leads to a change in the value of deceleration. This makes it possible to estimate the value of the changes in the state of the atmosphere depending on the latitude and time of day.

Calculations to determine the density of the atmosphere, taking into account the changes in the location of the perigee of the orbit, were made on the basis of observations of the first sputniks. The calculations showed that the product of the density and the square root of the "height of uniform atmosphere" increases as the orbit passes from the night side of the atmosphere to the day side and reaches its maximum at noon. An analysis of deceleration also revealed a decline of this value during the passage from the more northerly regions into those of the equator. Mention should also be made of the fact that there is good coincidence in the values of densities calculated on the basis of observations of Sputniks I and II and the carrier-rocket of Sputnik I.

The data obtained provide grounds for the conclusion that the temperature of the atmosphere at altitudes around 225 kilometers is higher than was formerly supposed on the basis of theoretical considerations. The discovery of higher temperatures of the atmosphere confronts geophysicists with the problem of the powerful sources of energy which heat the atmosphere. The known ultraviolet and X-ray radiation of the Sun is hardly sufficient for that. At present only various hypotheses can be advanced. It may be assumed, for example, that the upper atmosphere in the Arctic regions is intensively heated by solar radiation of minute particles. It is possible that in general the entire upper atmosphere is additionally heated either by infra-sound waves coming from the troposphere or by electric currents arising in the electrically-conductive ionized air as a result of its movement in the earth's magnetic field.

Continued study of the upper atmosphere with the help of rockets and sputniks will enable scientists to get the final answer to all these interesting and important questions.

Results of Ionospheric Exploration

Observation of the radio signals from the first sputniks has yielded new data on the outer ionosphere, *i.e.*, its regions starting at 300 to 400 kilometers above the earth. The ionosphere is the upper part of the atmosphere and contains a considerable quantity of free-charged particles (electrons and ions). Radio waves passing through the ionosphere are reflected, are partially or completely absorbed and their courses deflected. That is why radio methods have become the most effective means of exploring the upper strata of the atmosphere.

One of the basic parameters characterizing the condition of the ionosphere is the magnitude of electron concentration, *i.e.*, the content of free electrons in one cubic centimeter. Until now electron concentration was measured from the lower strata of the ionosphere to an altitude of 300 to 400 kilometers, where electron concentration has its so-called chief maximum.

These measurements were made mainly by ground ionosphere stations emitting short impulse radio signals of varied frequency and receiving their reflections from separate strata of the ionosphere.

As a result of regular measurements it has been established that the height of the chief maximum of the ionosphere and its electron concentration change from day to night, season to season, north to south, and from east to west. The greatest electron concentration observed over middle latitudes reached two million to three million electrons in a cubic centimeter, with electron concentration increasing at an average of 10 to 15 times from an altitude of 100 to 110 kilometers to an altitude of 300 to 400 kilometers.

It is highly important to know how electron concentration changes above the chief maximum, *i.e.*, in the outer ionosphere. This is necessary for, among other things, an understanding of the interaction of the Sun's ultraviolet radiation with the atmosphere, a study of the conditions of the propagation of radio waves and other processes taking place in the ionosphere. It is, however, impossible to explore the outer strata of the ionosphere by observing the radio signals reflected from them, because the radio waves emitted from the Earth are either completely reflected by the lower strata or pass into outer space. Some information on the outer ionosphere may be obtained by studying the radiations of the Sun and the stars reaching the Earth as well as the radio signals reflected from the Moon.

Observation of the propagation of radio waves of various frequencies emitted by the sputniks at different altitudes are a new means of exploring the outer ionosphere.

In receiving radio signals from the first sputniks at a frequency of 40 megacycles their "radio dusk" and "radio dawn" were fully observed in a number of cases and their respective time was recorded. In contrast to the optical dawn or dusk of a sputnik, which are characterized by the fact that at this moment the beam of light from the sputnik to the observer comes in a straight line, the radio beam at "radio dawn" or "radio dusk" is deflected in the ionosphere.

Because of this, "radio dusk" occurs later than optical dusk and conversely, "radio dawn" precedes optical dawn. The difference in time between optical dawn and "radio dawn" (or optical dusk and "radio dusk") makes it possible to determine the magnitude of the deflection of the radio beam. Since the deflection of the radio beam in the ionosphere depends on the change in electron concentration with altitude, it is possible, by assuming a certain law of the change of electron concentration, to calculate theoretically its magnitude at different altitudes. In doing so the influence of the lower strata of the ionosphere can be estimated on the basis of direct measurements carried out by a network of ground stations.

The data obtained from observation of the radio signals from the first sputniks make it possible to consider that electron concentration in the outer ionosphere (above the chief maximum) decreases with the rise of altitude 5 to 6 times slower than it increases below the maximum. Thus, from an altitude of 100 kilometers to an altitude of 300 kilometers the electron concentration mounted during the period of observation (in October) approximately tenfold, and from an altitude of 300 kilometers to 500 kilometers it dropped by half.

It should be noted that similar changes in electron concentration with the rise of altitude were also registered in launching a Soviet high-altitude rocket, which was reported in *Pravda*. In this experiment electron concentration at a height of 473 kilometers was of the order of 1,000,000 electrons in a cubic centimeter.

Study of Cosmic Rays

For studying cosmic radiation Sputnik II was equipped with two instruments for registering the number of particles of this radiation. Circling the Earth, the sputnik traveled at different distances from its surface. That is why measurements of the cosmic rays on the sputnik made it possible to ascertain the dependence of the number of particles on altitude. An analysis of the material obtained has revealed that the intensity of cosmic radiation increases by approximately 40 per cent from the minimal height of the orbit (225 kilometers) to an altitude of 700 kilometers. This increase is due primarily to the fact that the screening effect of the Earth diminishes as altitude increases, and the cosmic rays are able to reach the instrument from a great many directions.

The Earth's magnetic field also creates an obstacle to cosmic radiation reaching the earth. Deviation of the particles of cosmic beams in the Earth's magnetic field results in the fact that only particles whose energy exceeds a certain value can reach every point on the Earth's surface in a definite direction. Naturally, the farther away we go from the Earth, the weaker the magnetic field becomes, and the smaller is its effect upon the cosmic rays. Calculations show that the cosmic ray intensity increasing with altitude as measured in the flight of the sputnik can be explained by the above-stated reasons.

A study of cosmic rays through instruments installed in a sputnik may also reveal the dependence of the intensity of cosmic rays on latitude and longitude. This makes it possible to obtain new information about the Earth's magnetic field. Measurements of the magnetic field on the surface of the Earth give an idea about the character of terrestrial magnetism and allow to foretell what the magnetic field should be at great distances from the Earth. Proceeding from this it is possible to calculate the expected distribution of the intensity of cosmic rays over the Earth's surface. Specifically, it is possible to indicate the lines of the constant intensity of cosmic rays (isocosm). Measurements of cosmic rays made during the flight of the sputnik have shown that the lines of constant intensity obtained experimentally and calculated theoretically differ substantially. This result is in good agreement with the conclusion of the American physicist, Simpson, who organized a large series of flights of high-altitude aircraft over the equator. They showed that the equator determined by means of cosmic rays does not coincide with the geomagnetic equator.

Consequently, there is a considerable divergence between the characteristics of the Earth's magnetic field obtained by means of cosmic rays, on the one hand, and by measuring the magnetic field on the surface of the Earth, on the other. These divergences are due to the fact that the trajectories of cosmic rays are determined by the magnetic field at very high altitudes, while direct measurements characterize the magnetic field near the surface of the Earth. Cosmic rays make it possible to "sound" the earth's magnetic field at great distances from the Earth, permitting a new approach to the study of the Earth's magnetic field and the system of electric currents in the upper atmosphere.

Observation of cosmic rays with the aid of sputniks have made it possible also to register variations in the intensity of this radiation. These variations are, obviously, connected with the condition of the interplanetary environment near the earth. One instance of a sharp increase (by 50 per cent) in the number of particles of cosmic radiation was registered. At that time, however, ground stations did not detect any essential increase in the intensity of cosmic radiation, and this event is now being studied in detail. It is possible that it was caused by the sun's generation of particles of low-energy cosmic rays (which are strongly absorbed by the Earth's atmosphere) or by the sputnik passing through streams of high-energy electrons (connected with the minute-particle radiation of the Sun). Such phenomena could not be registered before, because the instruments for sustained observation of cosmic rays were situated only on the surface of the Earth. The sputniks for the first time made it possible to fully investigate primary cosmic radiation.

Biological Investigations

Over the past decade Soviet scientists have carried out a large number of experiments in the upper strata of the atmosphere. Test animals were sent up by rockets to altitudes of several hundred kilometers. The data obtained has brought us nearer to learning the nature of biological phenomena occurring in conditions closely resembling these of flying in outer space. Direct experiments to study the influence on a living organism of such factors, which cannot be reproduced on Earth, became possible. But only in the sputnik has it become possible to carry out biological experiments in space flight: primarily to study the influence on a living organism of a prolonged state of weightlessness, primary cosmic radiation, certain types of solar radiation, etc.

The data obtained in carrying out the program of medical and biological research in Sputnik II are of great value. As is well known, a test animal, the dog Laika, made a space flight in this sputnik.

Of great interest is the behavior and condition of the test animal in the most difficult, from the biological viewpoint, stage of the sputnik's flight—in its launching and entry into orbit. On its ascent to the orbit the sputnik traveled at an accelerated pace, the acceleration exceeding many times that of gravitation on the surface of the earth, and the seeming weight of the animal increasing with the acceleration.

During the ascent the animal was in a position for the acceleration to act on it in the direction from the chest to the back, which pressed the animal to the floor of the chamber. This position of the animal was chosen because it is a most favorable one for the organism. Simultaneously with acceleration, the vibration and noise of the rocket's engine reacted on the animal during the ascent.

The behavior and condition of the animal during the sputnik's ascent to the orbit was registered quite fully. The information obtained indicates that the animal withstood the increase in its seeming weight and continued to move its head and body freely only until a certain point of the acceleration. After that the animal was pressed to the floor of the chamber and no more or less noticeable movements were registered.

A study of the data obtained from the sputnik showed that immediately after the launching, the frequency of the heart contractions approximately trebled as compared with the initial frequency. The electrocardiograms have not revealed any morbid symptoms. They showed a typical picture of quickened heart-beat, the so-called sinus tachycardia. Later on when the effect of the acceleration not only continued but mounted, the heart-beat frequency diminished.

One can easily imagine that as the seeming weight of the animal increased, the respiratory movements of its thorax became difficult, breathing became more shallow and frequent. Indeed, telemetric recordings show that in the sputnik's ascent to its orbit, the animal breathed three to four times as fast as it did at the beginning.

There is reason to assume that the changes observed in the condition of the animal's physiological functions owe their origin to the sudden action on the organism of sufficiently strong external irritants: acceleration, noise and vibration, which began at the launching and continued on the ascent. An analysis of the data obtained and their comparison with the results of preceding laboratory experiments indicate that the animal withstood the flight quite well from the launching to the entry of the sputnik into orbit.

After the sputnik got into orbit, the centrifugal force acting upon it balanced the earth's attraction, and a state of weightlessness set in. In this condition the animal's body ceased to press upon the chamber's floor, and by contracting the muscles of its extremities it easily pushed itself off the floor. The recordings suggest that these movements were brief and rather smooth.

As the animal's thorax was no longer pressed under the influence of its increased weight, the frequency of its breathing declined. After a very brief period of quickened heart-beat, the systole frequency continued to diminish, consistently approached its initial level. It took, however, about three times as long for the number of heart beats to reach the initial level as it did in laboratory experiments in which the animal was subjected to the same acceleration as when the sputnik was put into orbit.

This is most probably connected with the fact that in the ground experiments the animal, after the acceleration ended, was in normal conditions, while in the sputnik the acceleration was replaced by a state of complete weightlessness.

In this state the animal's nerves whereby it feels the position of its body in space were not sufficiently affected by the external irritants. This conditioned the change in the functional state of the nervous system regulating blood circulation and respiration and determined a certain extension of the time for the normalization of these functions after the acceleration effect ended.

It is also possible that this phenomenon was somewhat intensified by the action of concomitant factors during the ascent—vibration and noise, which were greater than in the laboratory experiments.

It should be noted that the change in the physiological functions, registered in the animal at the beginning of the sputnik's movement along its orbit, coincides basically with the results of previous investigations with high-altitude rockets.

An analysis of an electrocardiogram recorded during the state of weightlessness revealed certain changes in the configuration of its elements and the duration of separate intervals. The observed changes were not of a pathological nature and were connected with the heightened functional activity during the period preceding the state of weightlessness. The electrocardiogram showed transient reflected nervous changes in the regulation of the heart's action. In the subsequent period the picture of the electrocardiogram grew increasingly closer to that characteristic of the animal's initial condition. In spite of the unusual state of weightlessness the animal's motions were moderate.

The normalization of blood circulation and respiration during the period of weightlessness, *i.e.*, during the period of the sputnik's movement along its orbit, evidently indicates that this factor in itself did not cause any essential and stable changes in the state of the animal's physiological functions. Thus it may be said that the animal well endured not only the sputnik's ascent to the orbit but also the conditions of travel along the orbit.

In ensuring the conditions necessary for the animal's vital activity in a prolonged flight in a sputnik, it is most important to provide a proper gas environment, the composition and pressure of which should not cause violations of the animal's physiological functions. This task could be accomplished only by the use of an hermetically sealed chamber in which normal atmospheric pressure with an oxygen content of 20 to 40 per cent and a carbon dioxide gas content of no more than one per cent was maintained by air regeneration.

Special highly active chemical compounds which, absorbing water vapors and carbon dioxide, emitted oxygen were used as regenerating substances. These chemical compounds absorbed also such noxious gases formed in the process of the animal's vital activity as ammonia, for example. An analysis of the data obtained showed that oxygen was emitted in sufficient quantities. The fact that the pressure in the chamber did not drop shows that it was effectively sealed.

We did not succeed in getting any definite idea about the impact of cosmic radiation upon the test animal. No clear physiological effect of its action was directly detected. A detailed study of this question requires thorough and prolonged investigations of the animal after flight, which is planned in further experiments.

The first appraisal of the results obtained shows most clearly that an animal endures the conditions of cosmic flight well. The positive—in this sense—result of the experiments makes it possible to continue and expand with still greater persistence investigations aimed at ensuring safety to the health and life of man in space travel.

Pravda, April 27, 1958

Sputnik III in Flight

Possibility of Launching Rockets into Outer Space

The whole world is following the flight of Soviet Sputnik III, launched on May 15, 1958, and placed in orbit by means of a powerful carrier-rocket on a predetermined trajectory of flight. After the carrier-rocket with its sputnik reached a speed of more than 8,000 meters per second, the sputnik was separated from the carrier-rocket by means of special devices and began circling the Earth in an elliptical orbit.

Soviet Sputnik III is much superior to its earlier predecessors. Two and a half times heavier than Sputnik II and sixteen times heavier than Sputnik I, it has a hermetic, cone-shaped body 3.57 meters long and a diameter of 1.73 meters, containing a large number of systems for carrying out the most complex scientific experiments, designed principally for the study of the phenomena taking place in the upper layers of the atmosphere and the influence of cosmic factors on processes occurring in the upper atmosphere.

The sputnik is equipped with efficient radiometric apparatus, ensuring an accurate measurement of its movement in orbit. The multichannel radiotelemetrical system of the sputnik possesses a high resolving power. It can transmit to the earth an extremely large amount of information on the scientific measurements carried out in the sputnik. The radiotelemetric system also includes a number of devices which continually "memorize" the data of the scientific measurements as the sputnik moves along its orbit. When the sputnik flies over ground measuring stations, the "memorized" information is transmitted from the sputnik at great speed.

The sputnik is equipped with a timer ensuring the automatic functioning of its scientific and measuring apparatus. This timer is composed entirely of semiconductors. There is also wide use of new semiconductor elements in all the measuring, scientific and radiotechnical apparatus—several thousand semiconductor units in all. The power for the apparatus is supplied by highly efficient electrochemical sources of energy and semiconductor silicic batteries transforming the solar rays into electric power. The solar batteries consist of a number of elements which represent thin plates of pure monocrystallic silicon with a pre-set electronic conductor. The tension created by separate silicon elements equals about 0.5 volts, and the coefficient of transformation of solar energy reaches 9-11 per cent. A corresponding combination of elements make it possible to obtain the necessary tension and value of the current. The solar battery on Sputnik III will enable scientists to study its performance in detail under conditions of cosmic flight.

The great weight of Soviet Sputnik III testifies to the high qualities of the carrier-rocket which placed it in orbit. It is the opinion of prominent scientists that only sputniks of great weight can promote the rapid solution of the problem of space flight which cannot be solved with the aid of small satellites having very limited possibilities for scientific research.

The continued increase in the weight of the Soviet sputniks bears evidence of the further possibilities of our rocket engineering. At present it is already possible to launch a rocket into outer space beyond the bounds of the Earth's gravitational force. For such a space rocket to have scientific significance and be a real step toward implementing interplanetary flight, it is necessary that it be properly equipped with scientific and measuring apparatus and that its launching yield new data on the physical phenomena in the universe and on conditions of space flight.

Scientific Station in Outer Space

The Soviet Sputnik III is an automatic scientific space station in the full sense of the word. Its arrangement and design have been greatly improved upon over those of the earlier sputniks. In designing the sputnik account was taken of a large number of specific requirements connected with the carrying out of diverse scientific experiments and the arrangement of a large number of scientific and measuring instruments in it. The possibility of individual scientific instruments affecting each other called for a thorough elaboration of the sputnik's layout and arrangement of the sensitive elements of its scientific apparatus.

The scientific apparatus in Soviet Sputnik III will make possible the extensive study of geophysical and physical problems. The structure of the ionosphere will be investigated by observing the propagation of radio waves emitted from the sputnik by a high-capacity radio transmitter. In addition the sputnik has apparatus for directly measuring the concentration of positive ions along its orbit. Special apparatus will make it possible to gauge the sputnik's own electrical charge and the electrostatical field in the layers of the atmosphere through which the sputnik passes. It is also measuring the density and pressure in the upper layers of the atmosphere. A mass spectrometer in the sputnik will make it possible to determine the ion spectrum which characterizes the chemical composition of the atmosphere.

To study the magnetic field of the Earth at high altitudes the sputnik has a self-orientating magnetometer which measures the full intensity of the magnetic field.

A number of experiments are devoted to a study of various radiations falling on the Earth and influencing major processes in the upper layers of the atmosphere. Also cosmic rays and the corpuscular radiation of the Sun is being studied with the help of the sputnik. Registration of the intensity of cosmic rays on almost the entire surface of the Earth will yield new data on cosmic radiation and the Earth's magnetic field at high altitudes. Experiments are being conducted to determine the quantity of heavy nuclei in cosmic radiation. Experiments on the corpuscular radiation of the Sun will shed new light on the nature of the ionosphere, aurorae, and other phenomena in the atmosphere. Several monitors will register micrometeor hits.

Highly important is a new experiment in registering the photons contained in cosmic radiation, an experiment which will make possible the obtaining of data on short-wave electromagnetic radiation in outer space. This is the first experiment making possible the study of cosmic radiation absorbed by the atmosphere, and the first step in opening up a new stage in astronomy—a study of phenomena in the universe through the short-wave radiation of stars. A number of experiments are designed to investigate flying conditions in outer space. These include a study of the thermal regime in the sputnik, the sputnik's orientation in space, etc.

An electronic timer automatically controls the work of the entire scientific and measuring apparatus, periodically switching it on and off. This timer also periodically transmits highly-accurate time signals necessary for the subsequent reduction of the results of scientific measurements of astronomical time and geographical coordinates. The fact that a large number of scientific investigations are to be carried out with the Soviet Sputnik III makes it a real cosmic research station. The creation of such a station on a high technical level and its equipment with a wide range of apparatus was made possible through the great size of the sputnik.

The Sputnik's Orbit and Observation of Its Movement

The sputnik will pass over all points of the globe between the Arctic and Antarctic Circles. This still further enhances the value of the scientific experiments conducted by the sputnik. The parameters of the sputnik's orbit have been chosen so as to ensure scientific investigations in the most interesting range of altitudes.

The Soviet Sputnik III was placed into elliptical orbit with an apogee of 1,880 kilometers. In the beginning the sputnik circled the earth 105.95 minutes, making about 14 revolutions in 24 hours. Later the circling period and the apogee of the orbit will gradually diminish because of the deceleration of the sputnik in the upper layers of the atmosphere. According to preliminary estimates Sputnik III will travel along its orbit much longer than the two previous Soviet sputniks.

The movement of Sputnik III in relation to the Earth is similar to the movement of the earlier Soviet sputniks. In the middle latitudes, due to the rotation of the Earth and the precession of the orbit, each successive spiral passes approximately 1,500 kilometers west of the preceding spiral. The speed of the orbit's precession is about 4 degrees in 24 hours.

The sputnik's movement is observed by radiotechnical and optical methods. The instruments and method of observing Sputnik III have been greatly improved.

The radar-obtained data on the coordinates of the sputnik are automatically reduced to the one given astronomical time. Then through special lines of communication these data are transmitted to a common coordinating computing center where the measurements received from various stations are automatically fed into rapid electronic computers which analyze all these data and calculate the orbit basic parameters. On the basis of these calculations, the further movement of the sputnik is predicted and its ephemerides are given.

This highly complex measuring set-up which includes a large number of electronic, radiotechnical and other devices ensures the measurement of the sputnik coordinates and the rapid determination of the parameters of its orbit much more accurately than the measurements of the movement of the previous sputniks.

Besides the above, DOSAAF clubs, radio direction-finder stations and a large number of radio hams are taking part in observing the sputnik. A 20.005-megacycle transmitter in the sputnik continually transmits radio signals in the form of 150- to 300-millisecond telegraphic messages. The power of the transmitter's radiation ensures the proper reception of its signals at large distances by ordinary amateur receivers. The systematic registration of these signals—particularly by tape-recording, which can easily be done by radio hams—will be of great scientific importance.

Of considerable interest are also radio observations of the sputnik's movement, based on the use of the Doppler effect. As observations of the earlier Soviet sputniks show, this method is highly effective and given a good reduction of the measurement results to astronomical time, makes possible the procurement of accurate data on the sputnik's movement.

In organizing optical observations of the movement of the Soviet Sputnik III, account has also been taken of the experience gained in watching the earlier sputniks. The network of ground stations for optical observation has been expanded and a number of foreign observation stations have been added to it, while the photographic observation methods have been notably improved.

Of special interest is the use of optical electronic transducers for photographing the sputnik, making it possible to obtain a clear photographic picture of it at a very great distance. Models of sputnik-photographing apparatus using optical electronic transducers were successfully tested in observing Sputnik II.

Exploration of the Ionosphere

A number of important characteristics of the ionosphere have been studied altogether inadequately.

A prominent place in the program of scientific research carried out through Sputnik III is given to exploration of the ionosphere. To study in detail the structure of the ionosphere and its basic characteristics is a major geophysical problem. It should be noted that the solution of this problem is of primary importance for ensuring reliable radio communication of the Earth with space rockets as well as accurate radio measurements connected with the flight of these rockets.

On Sputnik III, as was the case also with the first two sputniks, an extensive program of ground observation of the propagation of radio waves is being carried out. The results of these observations should yield extensive data on the state of the ionosphere.

In addition to ground measurements, direct measurements of the characteristics of the ionosphere are being taken by Soviet Sputnik III.

For measuring the concentration of positive ions along the orbit, two ionospherical screen traps have been installed on the surface of the sputnik. In Soviet Sputnik III, a mass spectrometer has been installed to determine the spectrum of the mass of positive ions contained in the Earth's ionosphere. When the ion mass number is known, it will be possible also to draw certain conclusions regarding the chemical composition of the ionosphere.

Measurement of positive ion concentration will make it possible for the first time to obtain data on the full concentration of charged particles in the ionosphere over various geographical areas of the Earth at different heights, as well as on concentration changes which occur when passing from the region illuminated by the Sun into the region of shade, and back. These data are highly important for an understanding of the processes of interaction of solar radiation with the Earth's atmosphere.

A comparison of measurements carried out in the region lying below the so-called chief maximum of ionization, at an altitude of 300 to 350 kilometers, with the results of observations by ground ionosphere stations, will enable scientists to reach a number of conclusions on the concentration of negative ions at these altitudes and on air ionization created by the movement of the sputnik itself.

It may be expected that the measurements of the concentration of positive ions will yield new data on the structure of the outer region of the ionosphere, supplementing the information on this region obtained by means of rockets and the earlier sputniks. One may likewise expect that the dimensions of ionosphere heterogeneities will be measured.

Utilization of sputniks for the study of such characteristics of the ionosphere as the concentration of ions and the spectra of their masses requires an understanding of the disturbances which the sputnik causes in the environment. Therefore, measurement of the electric charge of the sputnik which causes a redistribution of the charged particles near it is desirable also for greater accuracy of the results of these experiments. On the other hand, information about the electrical charge combined with data on the concentration of its ions can enable scientists to determine in a number of cases such a characteristc of the ionosphere, which is difficult to measure, as its temperature. The system of measurements adopted makes it possible to determine the value of the electrostatic field, and the use of two symmetrically-placed instruments of the electrostatic fluxmeter provides the possibility of determining not only the sputnik's own charge, but also the outside electrostatic field.

Measurements of a Geomagnetic Field

A magnetometer on a sputnik makes possible a magnetic survey along the entire globe in a short time, opening up exceptional possibilities for studying the variable part of the magnetic field.

According to present concepts, magnetic disturbances are caused by the movement of strong currents in the ionized layers of the atmosphere. Up to now only one direct experiment has been made with the help of a magnetometer installed in a rocket, testifying to the existence of such current systems.

The sputnik moving along its orbit will cross the ionized layers of the atmosphere many times. The existence of current systems can be seen from the fluctuations in the tensity of the magnetic field. Only by special methods of observation and analysis of the data could the part pertaining to the field of the supposed current systems be separated from the tensities of the field measured by the magnetometer. For this reason the program for investigating the spatial distribution of the constant part of the geomagnetic field and that of the field of variation in general cannot be combined in an experiment.

The main task of the experiment of Sputnik III is to study the spatial distribution of the constant geomagnetic field at high altitudes and compare the spatial distribution of lines of equal intensity of the magnetic field with the lines of equal intensity of the cosmic rays.

Measurement of the magnetic field from the sputnik entails considerable difficulty because of the fact that the position of the sputnik in relation to the vector of the geomagnetic field is constantly changing. The magnetometer requires high sensitivity for making a wide range of measurements; at the same time the measuring instruments of the magnetometer are affected by the magnetic parts of the other equipment.

A magnetometer capable of overcoming these difficulties has been installed on the sputnik. It is an apparatus whose measuring instrument is automatically oriented in the direction of the full vector of the geomagnetic field under any orientation of the sputnik. Compensating current serves as a measure of the magnetic field and its changes. It is passed through a coil set up on the measuring instrument in such a way that the current should fully compensate the Earth's field in the volume taken up by the measuring instrument.

Two potentiometers placed on the orientation section make it possible to determine the position of the body of the sputnik in relation to the Earth's field and the speed of the rotation of the sputnik on its axis.

New Methods of Studying the Universe

The study of cosmic radiation enables scientists to obtain data about the processes of the origin of particles possessing very high energies in the distant realms of cosmic space. Moving along its orbit, the sputnik provides the opportunity of registering separately the cosmic radiation of different energies. The cosmic-ray counter installed on the sputnik will make it possible to obtain new data on changes in the intensity and the energy spectra of cosmic radiation.

Of special significance is the search for the tiniest particles of light, photons, in cosmic rays. Photons possessing considerable energies, so-called gamma rays, can indicate the source of this radiation better than any other component of cosmic radiation.

The finding of gamma rays in cosmic radiation poses great difficulties, all the more so since at present it is impossible to predict their intensity. A sputnik, existing for a long time outside the Earth's atmosphere can provide exceptional opportunities for discovering this new component of cosmic rays.

The instrument installed on the sputnik makes it possible to try experimentally for the first time to find gamma rays in primary cosmic radiation. If this attempt is successful, it will be possible to speak of a new method of studying the universe.

Quite considerable data on the composition of primary cosmic rays has been obtained by sending instruments into the stratosphere on sounding balloons. But a number of data on the primary composition cannot be obtained by taking measurements in the stratosphere, since even the small layer of substance which always covers the instrument changes the composition of the cosmic rays. So far it is not known whether cosmic rays have a noticeable number of nuclei of elements heavier than iron. The installation on the sputnik of an instrument for registering nuclei of heavy elements makes it possible to answer this question of importance for science.

Artificial Earth satellites are an effective means for studying solar corpuscular radiation. The present time is particularly favorable for investigation of corpuscular radiation which has increased because of greater solar activity.

With the help of apparatus on Sputnik III it will be possible to obtain valuable material on the geographic, altitudinal and diurnal distribution of the corpuscular streams. Together with registering the corpuscular radiation of the sun, the equipment enables scientists to get additional data on solar X-ray radiation, which will also be registered by the corpuscular indicators. It will be possible to distinguish this radiation from corpuscular radiation by the direction from which it comes and the abscence of reflection from the Earth's atmosphere. Moreover, it can be distinguished by the time of its appearance, since corpuscular radiation spreads slower than the electromagnetic one.

The use of sputniks enables scientists to clarify and extend existing concepts of the atmosphere's structure. The long stay of the satellite at a high altitude and a comparison of the results from spiral to spiral makes it possible to analyze the data in detail and to rule out possible errors.

With sufficient accuracy it will also be possible to assess the diurnal and latitudinal variations of density and pressure at the altitudes at which the sputnik is traveling.

It is known that tiny hard particles, micrometeors, move in interplanetary space, as they enter burning up the Earth's atmosphere. At present these particles can be studied only with the help of apparatus carried by rockets and particularly by artificial earth satellites.

Study of interplanetary matter is of essential importance for astronomy, geophysics and astronautics, and also for the solution of problems of evolution and the origin of planetary systems, since it will make possible the solution of a number of essential problems for contemporary cosmogenic theories.

It is also very important to know exactly the general quantity of meteoric substance which falls on the Earth's surface during a definite period of time. It is necessary to take into account the effect of the blows of meteoric bodies on the outer shell of the rockets and artificial satellite and also on the instruments installed on them. Account should also be taken of the danger of sputniks, especially interplanetary rockets, clashing with larger particles. Although there is little probability of such a clash, it exists and it is important to be able to assess it properly.

Thus, the vast complex of interconnected investigations conducted in all directions with the help of this sputnik will greatly contribute to the development of science. The launching of the third Soviet artificial Earth satellite is one of the most remarkable events of the International Geophysical Year. The large size of the sputnik and the high degree of its automation bring Soviet science and technology closer to the creation of space ships.

Press Release, May 29, 1958

Design of the Third Soviet Sputnik

THE AIRTIGHT body of the third Soviet sputnik has a conic shape and is made of aluminum alloys. Its surface, like that of the first satellites, is polished, and was specially processed for the purpose of giving it essential values of the factors of radiation and absorption of solar rays. The removable rear bottom of the body is fastened to the butt frame with bolts. Special packing secures hermetic sealing of the joint. Before launching, the satellite is filled with gaseous nitrogen.

Inside the body of the satellite on the rear instrument frame, which is made of magnesium alloy, are arranged the radiotelemetering equipment, the radio apparatus for measuring the satellite coordinates, the program-timing device, the instruments of the system of thermocontrol and measurement of temperature, the automatic device which switches the instruments and chemical feed sources on and off. Also installed in the rear frame are instruments for measuring the intensity and composition of cosmic radiation and devices for recording the impacts of micrometeors. The frame is fixed to the power unit, which is on the shell of the body.

The major part of the instruments for scientific investigations are, together with power feed sources, also located inside the satellite body, on the other instrument frame which is in the forepart. On this frame are installed electronic sets of instruments that serve for measurement of pressure, ion composition of the atmosphere, concentrations of positive ions, the magnitude of electrical charge and intensity of the electrostatic field, intensity of magnetic field, intensity of corpuscular radiation of the Sun. Here also installed is the radio transmitter.

The position of sensitive elements (data pickup units) of the scientific equipment is determined by their purpose. In Fig. 38 the magnetometer (1) is located in the front part of the satellite with the object of maximum distance from the remaining instruments. The cosmic ray counters (4, 9, 10) are installed inside the satellite. Other data pickup units are placed outside the satellite's airtight body. The multiplier phototubes (2) which serve for recording corpuscular radiation of the Sun, are attached to the front part of the body. In cylindrical glasses, founded into the metal skin of the front part, are installed one magnetic and two ionization manometers (5) which measure pressure in the upper layers of the atmosphere. Near them are placed two electrostatic fluxmeters (7) which serve for measuring the electrical charge and intensity of the electrostatic field, and also the tube of a radio-frequency mass spectrometer (8) which determines the composition of ions at high altitudes.

On two tubular rods attached by swivel joint to the body shell, are mounted spherical reticular ion traps (6) which permit measuring the concentration of positive ions during satellite travel in orbit. In the arc of launching the satellite into orbit, the rods with traps are pressed in to the body surface. After the satellite is orbited, the rods are swiveled out.

In the rear bottom of the body are installed four data units for recording the impacts of micrometeors (11).

The transistor solar battery (3) is installed in the form of separate sections on the body surface. Four small sections are placed in the front bottom; four sections in the side surface and one section on the rear bottom. Such an arrangement of the solar battery sections secures its normal operation regardless of satellite orientation relative to the Sun.

The front part of the satellite is closed by a special protective cone which is jettisoned after the vehicle is launched in orbit. (See Fig. 39) The protective cone safeguards the satellite's front part, with its installations of data pickup units, from thermal and aerodynamic effects during passage of the carrier rocket through dense layers of the atmosphere. The cone consists of two half shells which divide, separating on jettisoning. Apart from the protective cone, a con-

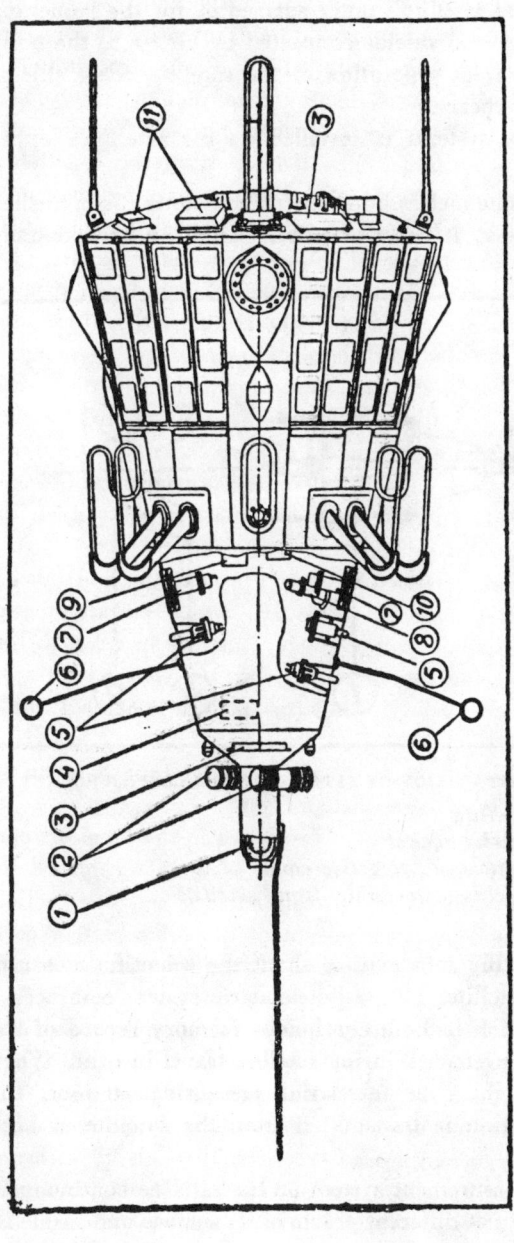

Fig. 38 SCHEMATIC DIAGRAM OF SPUTNIK III

1—magnetometer
2—multiplier photo tubes
3—solar batteries
4—cosmic ray photon recorder
5—magnetic and ionization manometers
6—ion traps
7—electrostatic fluxmeters
8—mass spectrometric tube
9—cosmic ray heavy nuclei recorder
10—device for measuring intensity of primary cosmic radiation
11—data units for recording micro-meteors

siderable portion of the satellite's outer surface is, for the launching arc, covered by four special shields connected by hinges to the body of the carrier rocket. With separation of the satellite, these shields remain on the carrier rocket.

A series of antenna systems is installed on the satellite's upper surface.

The satellite's multichannel radiotelemetric system is distinguished by high resolving power. It can transmit to Earth an exceptionally

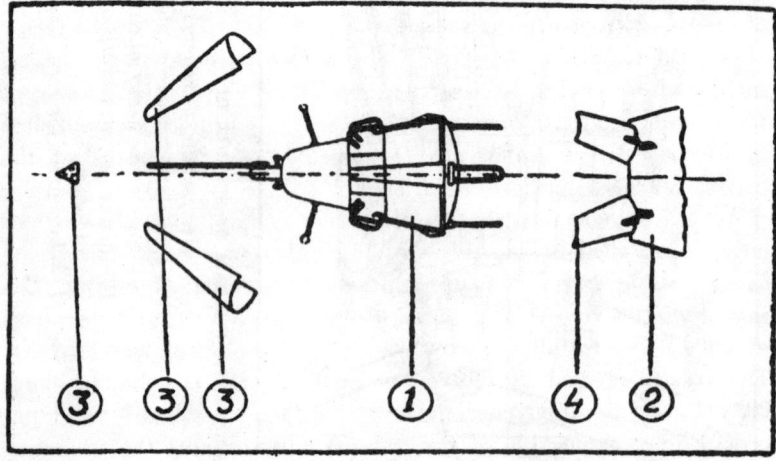

Fig. 39 DIAGRAM OF SEPARATION OF SATELLITE FROM CARRIER ROCKET

1—satellite
2—carrier rocket
3—detaching protective cone
4—shields detachable from satellite

large volume of scientific information about the scientific measurements made on the satellite. The radiotelemetric system embraces a number of devices which make a continuous memory record of the data of scientific measurements during satellite travel in orbit. When the satellite is in flight over terrestrial measuring stations, the "memorized" information is transmitted from the satellite at high speed.

The temperature measurement system on the satellite continuously records the temperature at different points of its surface and inside it.

An electronic program-timing unit accomplishes automatic control of the operation of all scientific and measuring instruments, periodically switching them on and off. This unit also periodically issues with great accuracy time marks, which is necessary for subsequent linkage of the results of scientific measurments to astronomical time and geographic coordinates.

A stable temperature regime on the satellite is secured by a system of thermocontrol which is a considerable improvement over the thermocontrol systems on the first satellites. Regulation of thermal conditions is achieved by means of variation of forced circulation of gaseous nitrogen in the vehicle and also by variation of the factor of natural radiation of its surface. For this purpose on the lateral surface of the satellite are installed regulating louvers that consist of 16 separate sections. Opening and closing of them is accomplished by electric drives which are regulated by the apparatus of the thermocontrol system.

Apart from chemical sources of electric current, a set of solar batteries is installed on the third satellite. These batteries convert the energy of the Sun's radiation directly into electrical energy. The solar batteries consist of a series of elements, which are thin plates of pure monocrystallic silicon with a preset electron conductivity. The voltage created by the separate silicon cells is equal to about 0.5 volt, and the solar energy conversion factor reached 9 to 11 per cent. Proper connection of the cells permits deriving the necessary voltages and current magnitude.

The launching of the third Soviet artificial earth satellite is a new evidence of the achievements of rocket propulsion in the Soviet Union, and one of the most remarkable events of the International Geophysical Year. The large size of the satellite, and the high degree of its automation brings Soviet science and technology closer to the creation of space ships.

Izvestia, May 20, 1958

Solar Battery on Sputnik III

by M. S. SOMINSKY
*Vice-director of the Institute of Semi-Conductors
at the Academy of Sciences, USSR*

The solar battery (Fig. 40) installed on Sputnik III for generating electricity to power the satellite instruments is in addition to electrochemical energy sources. Its working principle in conversion of solar energy to electricity was described by scientist M. S. Sominsky as follows:

The problem of direct conversion of solar energy to electricity has engaged the minds of scientists, technicians and inventors of various countries for many years. A quarter-century ago, the prominent Soviet scientists Academician A. F. Ioffe advanced a series of fruitful ideas which contributed to the creation of a semi-conductor photoelectric cell with high efficiency. At that time few believed that such a device was realizable in practice.

In recent years intense research work of physicists has been crowned with success. A new semi-conductor photoelectric cell was created with an efficiency of more than 10 per cent.

What is a semi-conductor photoelectric cell that converts the energy of radiation into electrical energy?

In essence it is a miniature solar electric station, for which absorbed radiant energy serves as "heat."

In fact, everyone knows that any plant "absorbs" solar energy with its leaves, and by means of this absorption creates the conditions for its growth. To the point, botanists have estimated that in combustion of wood the energy derived is at least 100 times less than the amount which the timber received during its life from the Sun. Translated into technical language, this means that efficiency here amounts to less than one per cent. The plant absorbs less than

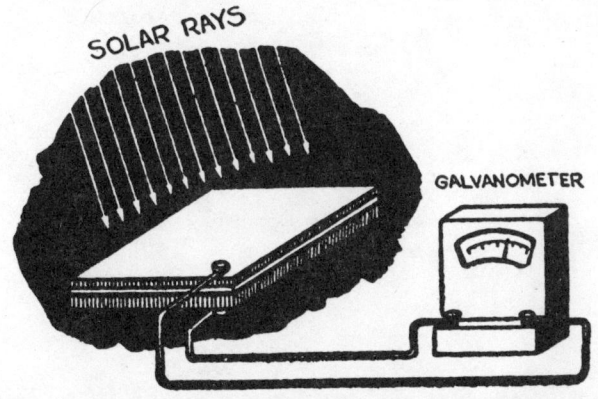

Fig. 40 Schematic diagram of a semi-conductor photo-electric cell, the so-called solar battery. Rays of light falling on the plate excite electron travel in a closed circuit. The galvanometer arrow deflects, showing the presence of current. The electric current developed in a powerful photo-electric cell can feed energy to a radio receiver or transmitter.

one per cent of the solar energy falling on the leaves. After multiple conversions of the solar energy—from radiant to chemical, from chemical to thermal, from thermal to mechanical, from mechanical to electrical—we derive only a fraction of a per cent of the energy. Such is the difficult and extremely long path which solar energy follows to reach any kind of motor, if ordinary fuel is used, for example.

Now the transistor "solar battery" converts the energy of solar rays directly into electricity. In particular, the new semi-conductor photo-electric cell of silicon is used as a highly efficient converter of solar energy to electrical. If individual photoelectric cells are connected one to another in series and in parallel then a "solar battery" is developed. The service term of this batery can be extremely long. Such a battery can be used for charging storage cells, powering rural telephone exchanges, radio receivers, various transmitting equipment and so forth. The exploration of cosmic space and launching of artificial earth satellites have opened up great possibilities for the "solar battery."

Izvestia, May 16, 1958

Model Comparison

by A. STERNFELD

A COMPARISON of Soviet and American satellites from the viewpoint of the energy which they possess after being put in orbit leads to a curious and demonstrative result. (The energy required for launching is considerably greater.)

The idea which lies at the basis of these calculations is simple. In order to put an artificial earth satellite in orbit, two conditions have to be met: first it is necessary to raise the satellite to a pre-set altitude and then to communicate to it the velocity required for circling around the Earth. Consequently, the mechanical energy which the satellite possesses in flight is equal to the sum of the work expended on its lift into space and the energy of its travel. For more exact calculation, it is necessary to consider that because the force of gravitation declines with elevation, the energy which must be expended in raising the satellite to its orbit is not proportional to elevation, as we assume in solving our "earth" problems. Another circumstance which has to be kept in mind is the ellipticity of orbits of satellites, by virtue of which the altitude of their flight varies. This, however, also can be taken into account.

Omitting the calculations themselves, let us see what conclusions can be drawn. If the complete mechanical energy of the first Soviet satellite be provisionally taken as equal to 100 units, then the energy of the second Soviet satellite will be measured by 633 units, and the third by 1,671 units. As for the American satellites, the full mechanical energy of the first and the third satellites (Explorer I and

Fig. 41

Explorer III) will be expressed in 18.2 units, and the second (Vanguard) in 2.1 units overall.

In order to give a graphic picture of the absolute magnitude of the energy here under discussion, we point out that the first Soviet satellite has an energy equivalent in magnitude to 10 trains of 1,150 tons each travelling at a speed of 80 kilometers an hour.

The magnitudes of the energies of the satellites are compared more perceptibly in Fig. 41. In the scale we have adopted, the large square which bounds the whole drawing corresponds to the energy of the third Soviet satellite.

Izvestia, May 18, 1958

Index

air-drag, 65, 97, 101
aerodynamic braking, 64, 99, 104, 119, 167
air regeneration, 32, 38, 39
air force, Russian, 109
Alexander the Great, 15
Aleksandrov, V., 196
Ambartsumyan, V., 155
Ananoff, A., 17
angular velocity, 64, 68-70, 78
animal experiments, 13, 44, 45, 57, 178, 186
antioverload suit, 42, 185
apogee, 65
astronautical speed, 21, 23, 24
Astrophysical Institute of Kazakh Academy of Sciences, 190
artificial Sun, 204-205
artificial gravity, 32, 51
artificial meteors, 92
artificial satellites, design of, 29, 30, 31
   of planets, 124-134
atmosphere, structure of, 56, 155-156, 163-165
   exploration of, 88, 92, 230
atomic engines, 28
atomic space ship, 35, 150
atomic fuel, 28, 58
aurora polaris, 92

Baklund, O. A., 130
benzine, 26
Belyaev, A., 16
Berg, A. I., 206
Bergerac, Cyrano de, 16
biology of space flight, 91, 177, 180-188

248 Index

Bogdanov, A., 16
Brazil, 67
Burgess, A., 17

carrier rocket, 31, 64, 74, 160, 168, 226
Chinese legend, 15
Chkalov Central Aeroclub, 14
circuit or orbit period, 63
circular speed, 21, 61
Clarke, A., 17
comets, 134-136
Cooper, J. C., 147
Copernicus, N., 16
cosmic dust, 91, 160, 234
cosmic rays, 30, 57, 188
   sputnik study of, 175, 219-221, 232-233
cosmic speed, 21, 23, 55
Crocco, 146

Deimos (Mars satellite), 99, 111
dogs in space, 13, 30, 44, 171
day-night cycle, 81-83
Dobronravov, V., 159
Doppler effect, 171, 213, 229
DOSAAF, 169, 229

Earth, 19, 21, 61, 80, 85-87
   penumbra and umbra of, 81-82
Earth-to-Mars ship, 35
Earth-to-Venus route, 140-141
Edinburgh Observatory, 190
Equator, 67
escape velocity, 23
Esnault-Pelterie, R., 17, 51
Europe, 67
exhaust velocity, 28, 110

flight trajectory, 101-103

Gartman, H., 17

geophysical problems, 163
Gerathewohl, S. I., 43-44
Goddard, R. H., 17
gravity, 19, 22, 32, 41, 50, 101
Greek mythology, 15

Gremin, V., 200
Haley, A. G., 17, 143
Hester, A. W. B., 147-148
hydrogen, 26
Hurford, 146

ICBM, 199, 212
Indian epic, 15
Indonesia, 67
International Geophysical Year, 14, 121, 175, 211, 234
International Astronautical Congress, 43
International Astronautical Federation, 14
interplanetary space station, 30, 94-98
instrumentation, 29, 68, 171, 172
Ioffe, A. F., 240
ionosphere, 89, 170, 217, 230
Isakov, P., 180

Journal of British Interplanetary Society, 147
Jupiter, 19, 55, 99, 103
  trip to, 117-119
  satellites of, 128-134

Kazantsev, A., 192
Kaznevsky, V., 202
Kepler, J., 16
kerosene, 26
Kibalchich, N. I., 17
Kiseleva, T. P., 74
Kola Peninsula, 80
Kondratyuk, Y. V., 17
Kondratyuk, Yu. K., 34
Konstantinov, K. I., 17
Kukarkin, B., 18
Kulebakin, V. S., 205

Laika (first dog in space), 13, 30, 46, 171, 177, 185, 207, 221-224
Ley, W., 17
liquid food, 45-47
liquid fuel rocket, 17, 25, 26
life-saving gear in space, 60
Lomonosov, M., 114

magnetism, terrestrial, 90, 164, 227, 231-232
Mandl, V., 143
man in space, 40-60, 151
Mars, 13, 18, 55, 94, 99
   trip to, 109
   artificial satellite of, 128-134
mass, 55, 80
Mercury, 13, 19
   trip to, 119
   artificial satellite of, 128-134
meteorites, 35, 54-55, 91-92
meteor danger, 35, 54-55, 56, 91-92, 234
Mikhailov, O., 17
microatmosphere, 38, 39, 59
Moon, 15, 18, 21, 66-68, 94, 98-99
   artificial satellite of, 122-128
moon flights, 13, 16, 36, 49, 55, 103, 106-109, 136-139, 150, 161, 162
"Moskva-sputnik," 206
multistage rockets, 17, 18, 27, 32

Neptune, 19
   trip to, 120
   artificial satellite of, 128-134
Newcomb, S. 130
Newton, I., 16
nitric acid, 26
North Pole, 68, 74, 80-83
nutrition in space travel, 45, 53-54

Oberth, H., 17
Observatory of Kharkov University, 19
Obukhov, A., 163

optical tracking, 63, 68, 69, 70-74, 78, 169, 189-191, 214, 229
orbital speed, 65
orbits, of earth satellites, 61-75, 93-94
   of space ships, 121-134
overload, 41
oxygen deficiency, 38-39, 51, 188
oxygen, liquid, 26, 39
ozone, 56

parabolic trajectory, 112
parabolic velocity, 23, 118
perchloric acid, 26
perigee, 65
Phobos (Mars satellite), 99, 111
Pluto, 19, 100; trip to, 120, 128-134
Pokrovsky, A. V., 44
Pokrovsky, G. I., 199
Potsdam Observatory, 190
Pravda, 158, 224
Promyshlenno-Ekonomicheskaya Gazeta, 162
Prozorovsky, Yu. N., 195
Pulkovo Observatory, 190
Purple Mountain Observatory, 190

radar, 68
radiation, 28, 56-57
   solar, 157, 173-175, 207-208, 227, 233
   short-wave of stars, 228
Radio (journal), 195, 210
radio-tracking, 68, 170-171
radiowave propagation, 89-90
radio signals from satellites, 29, 68, 75, 170, 213
reaction motion, 25
retro-rocket, 104, 119, 203
rockets, 16, 17, 25-27
rocket aircraft, 17, 104-105, 196-201
rocket engines, 24, 26, 36, 51, 70, 102, 200, 201
rocket launching sites, 21, 36
rocket fuels, 25, 36, 101, 201

rocket guidance, 25, 36, 55, 58, 70, 92, 161-162
rocket propulsion elements, 25-36, 201
"Rocket Research Establishment," 17
Romick, 32

Saenger, E., 17
satellite observations, 67-71, 79, 84-93
satellite orbits, planetary, 19
   solar system, 121-143
satellite space station, 32-36, 66-68
Saturn, 19, 99, 120, 128-134
Schachter, O., 143
semi-elliptical trajectories, 112, 133
sidereal time, 61-62
simulated space flight, 59, 60
solar battery, 240-242
Sominsky, M. S., 240
South Pole, 68, 74, 183
space city, 32-33
space law, 143-148
space ships, 19, 24-33
   observatory, 84, 237
   interplanetary orbital, 136-143, 202-204
space vegetation, 32
Sputnik I, 13, 18, 24, 29, 63-65, 71-78, 149, 155, 159, 171, 211
   instruments of, 29, 160
Sputnik II, 18, 29, 46, 63-65, 71-74, 78, 84, 149, 166, 171, 211
   instruments of, 30, 166, 171
Sputnik III, 225-234
   design of, 236-239
   instruments of, 225-226, 235-239
sputnik energy in orbit, 243-245
sputnik findings
   on animal experiments, 51, 221-224
   on atmosphere, 214-217
   on cosmic rays, 219
   on ionosphere, 217-218
   on magnetism, 220
   on radiowave propagation, 170-171, 192-195, 208

Stemmer, Y., 17
Sternfeld, Ari, 13-153, 243-245
stratosphere, 56
Sun, 13, 19, 31, 57, 79-83, 95, 103
   artificial satellites of, 124, 128-134, 137

thermochemical fuel, 35
Tikhonravov, M. K., 17
Tikhov, G., 111
Tisserand, F., 132
Tolstoi, A. 16
Tombaugh, C. W., 100
training of astronauts, 59-60
trajectories, 136-143
Tsander, F. A., 17
Tsiolkovsky, K. E., 13, 16, 17, 32, 34, 51, 91, 95, 162

UNO, 147
ultraviolet rays, 31, 57, 157, 173
Uranus, 19
   trip to, 120, 128-134
USSR Academy of Sciences, Astronomical Council, 189
   Joint Commission on Space Travel, 13

Venus, 13, 18, 19, 94, 113
   trip to, 114-117, 128-134
Verne, Jules, 16

water regeneration, 53
weightlessness, 32-33, 39, 42-47, 161-162, 186
Wells, H. G., 16
Wilkins, John, 16

X-rays, 57

Zero gravity, 32-33, 42-44, 46, 47, 186
zodiacal light, 92